New Coastal Navigation

DAVID NICOLLE
RYA Yachtmaster Instructor

New Coastal Navigation

STANFORD
MARITIME

Published in 1987 by Stanford Maritime Limited
Member company of the George Philip Group
12–14 Long Acre, London WC2E 9LP

British Library Cataloguing in Publication Data

Nicolle, David
    New Coastal Navigation,
    1. Coastwise navigation
    I. Title II. Series
    623.89′29   VK555

    ISBN 0–540–07325–3

Printed in Great Britain by
Butler and Tanner Limited

Contents

Introduction

This book aims to cover the coastal navigation aspects of the RYA Yachtmaster's syllabus. Topics that require little, if any, conceptual understanding, such as buoyage or chart symbols, are not included. Subjects like these are well covered by the various nautical almanacs or textbooks.

In order to work the exercises given in this book, the reader will need plotting instruments – namely dividers, parallel rule and/or a suitable plotter (protractor) – as well as a simple electronic calculator. Other useful additions include an almanac, an Admiralty chart of your own sailing area, and Admiralty publication *Symbols and Abbreviations used on Admiralty Charts* (NP 5011).

My own background includes about thirty years' sailing and navigating in and around the Channel Islands, and it is common knowledge that this is an excellent area in which to serve one's navigational apprenticeship. For this reason many of the exercises are based on the Channel Islands, where strong tides and numerous rocks make navigation challenging and interesting to both pupil and instructor alike.

The book refers to two excellent charts of great contrast. One is a privately published, medium scale, non-metric chart of the Channel Islands – Stanford chart no. 16, published by Barnacle Marine Ltd – while the other is an Admiralty metric, large scale, instructional chart (BA 5043) of the Goodwin Sands. In this way, a wide variety of experience is provided. The first is included, the second must be bought.

The main teaching method is through examples and many exercises. The answers, many with explanation and working, will be found at the end of the relevant chapter. Most questions embody a new step or slant on a particular topic and it is, therefore, necessary to work patiently through most, if not all, of the exercises. Many readers will find benefit from re-reading and re-working some of the less familiar topics. It is best for the whole question to be read through first, rather than making a start piecemeal: in this way the direction of the question will be sensed and possibly some additional hint picked up.

Lastly, a few observations on the differences between practice navigation at home, or in the classroom, and the real thing. Many students feel discouraged when faced with their mistakes in the classroom, and imagine that they could never cope with navigating a yacht at sea. This is not necessarily the case. Classroom navigation is, in many ways, harder because all the data is artificial: although realistic, it is unreal. It is all too easy, in the middle of winter, to look up a June tide when the month specified was July, or to read a low water time for a high water time. At sea you are far more likely to spot such mistakes. On the other hand, it must be said that working conditions in the classroom are better; there is more space, more light and the desk top is less likely to be leaping about! There is usually less anxiety over the passage of time or extraneous distractions when working on a desk rather than a chart table. In reality, it is all too common when busily plotting a crucial fix, to be told that the spinnaker has wrapped itself around the forestay or that a collision situation has arisen or that someone can 'smell burning'. Good luck!

David Nicolle
Guernsey

Definition of Position

Latitude and longitude: the basic framework

As one delightful old Guernseyman once said to me, 'At sea, you've got to know where you're to', and although this is not the whole story, it certainly helps!

The least ambiguous way of defining position is by stating your latitude and longitude. The latitude of your position is its angular distance north or south of the equator, and the longitude is its angular distance east or west of the prime (Greenwich) meridian. A good knowledge of this fundamental system is essential in that it:

a Gives an unambiguous way of defining position.

b Makes for a fuller understanding of the charts you are using.

c Is essential for the fluent use of electronic position fixing systems, such as satellite receivers or Decca.

The globe

Although the earth is not exactly a sphere, we consider it as such for navigational purposes. A great circle is defined as the intersection of the surface of a sphere with a plane passing through its centre. The shortest distance between two points on the earth's surface is along the great circle which passes through them, an important fact when sailing long distances.

Latitude

At right angles to the earth's axis and dividing the globe exactly in half, we have the basic reference line for latitude — the earth's equator. To measure latitude north or south of the equator, we must place ourselves hypothetically at the earth's centre. Directly above us we would have the North pole, latitude 90°N, and below our feet the South pole, latitude 90°S. The British Isles lie nearer to the North pole than to the equator, their very approximate latitude being 55°N. We call lines of latitude parallels because of their basic relationship parallel to the equator. Except for the equator, parallels of latitude do not bisect the globe because they are not great circles. They do, however, provide the means of determining distance at sea, because by definition a nautical mile (abbreviated to M or n mile) is the distance of one minute of latitude at the place concerned. Since the earth is not a perfect sphere, the nautical mile (sometimes called a sea mile) varies in length from 6046 ft (1843 m) at the equator to 6108 ft (1862 m) at the poles. Navigation instruments such as logs and radar sets are calibrated on a mean figure of 6076 ft (1852 m) which has been adopted as the International Nautical Mile.

If one minute (1') of latitude equals one nautical mile, it follows that one degree of latitude is equal to 60 nautical miles (60 M or 60 n miles).

Longitude

The basic reference point for longitude is Greenwich, and lines of longitude are known as meridians. The Greenwich meridian runs from the North pole, through Greenwich, to the South pole, cutting the equator at right angles. Each meridian is a semi-great circle through the poles, and its angular distance from the Greenwich meridian is measured from the earth's centre as an angle in the same plane as the equator. Longitude is stated East (positive) or West (negative) of the Greenwich meridian. The words 'positive' and 'negative' refer to the manner in which longitude is keyed into certain types of electronic navigator. Note also, that in the southern hemisphere, southern latitudes are also entered negatively in using such equipment. The meridians, converging on the North and South poles, are our basic reference lines for direction. Thus latitude and longitude give the navigator his essential tools of distance and direction. This concept is fundamental to the understanding of all navigational principles.

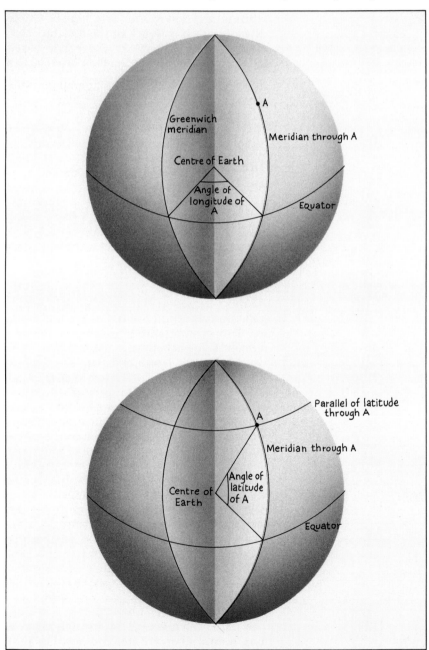

Note that the angle of longitude is measured at the centre of the earth in the plane of the equator. Longitude is expressed as east or west of the Greenwich meridian — it cannot be greater than 180°.

Charts and scales

Charts have to serve different purposes — for passage making, or for providing the more detailed information required at the entrance of a harbour. Hence they are drawn to different scales. Small scale charts (say, 1:500,000) are used for longer coastlines, while large scale charts (1:15,000 for example) are drawn to show the details of harbours and their approaches. The scale is the relation between a distance as portrayed on the chart and the actual distance on land or sea. Thus 1: 15,000 means that 1 ft on the chart is actually 15,000 ft on land or sea. While a great deal of detail can be shown on a large scale (harbour) chart, much of it has to be omitted on a small scale chart covering that area. Hence it is important always to use the largest scale chart available.

Differences between globe and chart

All charts and maps have to attempt the impossible task of representing the doubly curved surface of the earth on a flat piece of paper. Obviously this becomes more difficult the larger the area of the earth's surface the chart attempts to represent. This dilemma was successfully resolved by the Dutch cartographer Mercator. He realised that the basic need of the navigator was to reproduce angles and direction on the chart as they are on the surface of the globe, and for this reason he drew the meridians as equally spaced parallel lines. This means that a straight line between two points, cutting all the meridians at the same angle, is a straight line on the chart (called a rhumb line), and that angles on the earth's surface equal corresponding angles on the chart. But Mercator's projection has some disadvantages too: the distance scale varies with latitude (as we have seen above), land masses near the poles are greatly distorted, and the shortest distance between two points (which is the great circle through them) usually appears as a curve on the chart and is inconvenient for long distance sailing.

Subdivisions

In effect, a Mercator chart of the northern hemisphere exaggerates area in its northern limits and understates them in its southern limits. Only at a specified latitude is the scale exact. For this reason, all except the largest scale charts (usually 1:25,000) have no constant scale. Therefore, when working with medium or small scale charts, we must scale our distances in terms of minutes of latitude **in the latitude at which we are working**. This is the one serious penalty to pay when working on a Mercator projection of medium or small scale. On large-scale charts this enlargement is so slight as to be ignored.

Although nearly all navigational charts are based on Mercator's projection, gnomonic charts may also be met. A gnomonic chart is a projection of the earth's surface from the centre onto a tangent plane, and an important feature is that great circles appear as straight lines. Apart from ocean passages, gnomonic charts are used for polar regions and sometimes for large scale plans.

One degree of latitude or longitude represents a considerable distance. At the equator, for example, 1° of latitude and 1° of longitude enclose an area of 60×60 M or 3600 sq. miles. We must, therefore, have a suitable subdivision. Each degree is divided into 60 equal parts called minutes, and can be further subdivided into 60 seconds. The latter, however, in UK coastal navigation practice, is falling rapidly into disuse in favour of tenths and hundredths of a minute: i.e. minutes are now decimalised.

It is important not to confuse seconds of arc with tenths and hundredths of a minute, and to be able to convert readily from one to the other. For example:

52°13′39″ = 52°13′.65 (dividing 39″ by 60, to get 0′.65)
or 02°26′.42″ = 02°26′25″ (multiplying 0′.42″ by 60, to get 25″)

Those whose mental arithmetic does not cope with these simple calculations can use tables which are available, or a calculator.

A summary of latitude

1 Lines of latitude are called parallels.

2 Each degree of latitude is equal to 60 nautical miles, or 1° = 60 M.

3 Each minute of latitude is equal to 1 nautical mile, or 1′ = 1 M.

4 The parallels are regularly spaced on the globe.

5 The parallels are increasingly spaced on the chart (going north in the northern hemisphere, i.e. away from the equator).

6 The parallels are the basic means of acquiring distances.

7 The parallels are not great circles, with the exception of the equator.

8 The parallels are at right angles to the earth's axis.

A summary of longitude

1 Lines of longitude are called meridians.

2 The reference line is the Greenwich meridian.

3 Meridians are drawn parallel and equally spaced on Mercator charts.

4 Meridians are converging towards poles on the globe.

5 Meridians are the basic means of acquiring direction.

6 **Meridians must never be used to measure distances.**

Chart table instruments

Before starting to draw lines on a chart it is necessary to understand the instruments concerned. Most people are familiar with the use of dividers from geometry at school, but at sea dividers should be the type that can be operated in one hand.

Parallel rulers are used for transferring bearings and courses to and from the compass rose (a circle on the chart graduated into 360° and showing the cardinal points). They come in two types, the roller variety which is used in larger vessels, and the type with two parallel, hinged legs that can be walked across the chart and is better for a yacht. Even more handy is one of the various protractors or plotters that are on the market. A simple example is the Douglas protractor which consists of a perspex square graduated round the edge from 0°−360° in both directions, and with a hole in the centre. The square is also engraved with a grid, one line of which can be placed over a convenient meridian or parallel of latitude to measure a bearing or course on the chart. This simple device can be used for a number of plotting exercises on the chart table.

Practice exercises in latitude and longitude, and measuring distances and directions

Difficulties you may encounter
1 The number of degrees is usually given mid-margin of the chart if an exact meridian or parallel does not fall on the charted area.
2 Should your local area be west of Greenwich, then you must become used to scaling longitude from **right to left**. Proficiency in this is of great use when programming a Decca type navigator with a series

of positions, known as waypoints. Try to develop both quickness and, above all, accuracy.

Exercise 1

Recording positions in terms of latitude and longitude. (This exercise is self-checking.)

Using the Admiralty Instructional Chart, namely South Foreland to South Falls Head BA 5043, find an example of a tidal diamond (◇) These are purple in colour and there are eight examples to be found in various positions. Measure the position of the first diamond you find and compare your answer with that given on the top of the table of tidal streams in the NE extremity of the chart. For quickness use only dividers, and work to the centre of each diamond. It is important to write your answers down before checking them, in order to become familiar with the now accepted procedure. This is to write latitude first, placing the stroke denoting minutes above and slightly to the left of the decimal point, e.g. 51°06'.6N. Repeat this exercise using an Admiralty chart of your own area.

Exercise 2

Plotting latitudes and longitudes, using both practice charts

Plot latitude first using a parallel rule, placing a short line in approximately the correct longitude. Plot the longitude using dividers.

2a (Chart 5043) Plot 51°13'.05N, 1°36'.30E

2b Plot 51°21'.40N, 1°37'.90E

2c Note the true bearing (as shown on the outer circle of the compass rose) and distance *from* the South Goodwin Light Vessel *to* the East Goodwin Light Vessel, working from the small circle shown on each vessel's waterline.

2d Plot 51°09'.5N, 1°25'.40E. What is the true bearing and distance of this position from the East Goodwin Light Vessel?

2e Plot position 51°07'.00N, 1°29'.45E.

2f Write this position (2e) as a true bearing and distance from the South Goodwin Light Vessel.

2g Measure the distance between the South and North Goodwin Light Vessels.

Exercise 3

Using the enclosed Stanford chart no. 9 of the Channel Islands:

3a Plot 49°27'.2N, 2°31'.5W

3b Plot 49°01'.8N, 2°48'.3W

3c What is the true bearing and distance of position 3b to Noire Pute? This rock lies between Herm and Sark.

Note on distances The nautical mile (abbreciated M) is subdivided into tenths called cables. Half a mile can equally be referred to as 5 cables. For practical purposes a nautical mile is regarded as 6000 ft, and hence a cable is 600 ft. When we write a position as in Exercise 1 to 1/100th of a minute of latitude, we are working to a theoretical accuracy of a 1/10th of a cable or 60 ft. This is the *readability* of most Decca-type navigators. Their *accuracy* is one order greater than this, or about 1 cable (1/10th of a nautical mile or 0'.1 of latitude).

Speed The nautical unit of speed is the **knot**, defined as one nautical mile per hour.

13

To measure the total distance sailed around the island

Place the dividers on points A and B. Swing dividers, centre B, to point T, such that BT is the reciprocal of the next heading BC. Extend the dividers to span TC. With centre C, swing dividers to point Z, such that CZ is the reciprocal of the next heading CA. Extend the dividers to span ZA. The span ZA is the total distance sailed along the route ABCA.

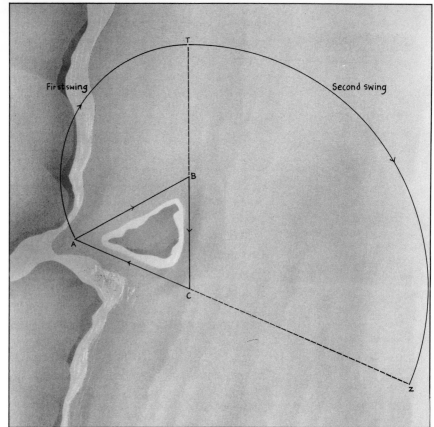

Handy hint: 6 knots is a convenient speed mathematically because the vessel proceeds at 1 cable per minute. This is a useful speed of advance when dead reckoning in poor visibility.

Formulae   Speed (in knots) $= \dfrac{\text{Distance (in M)}}{\text{Time (in hours)}}$

or $S = \dfrac{D}{T}$ also $D = S \times T$ and $T = \dfrac{D}{S}$

Example a   What is the speed of a vessel that sails 37.4 M in 6 hr 18 min? 6 hr 18 min = 6.3 hr, so speed = 37.4/6.3 = 5.94 knots. *Note:* to convert minutes to a decimal of an hour, divide minutes by 60. *Example:* 37 min = 37/60 hr or 0.62 hr.

Example b   How long will a 17.9 M journey take at 6.4 knots?

$\text{time} = \dfrac{\text{distance}}{\text{speed}}$ $\text{time} = \dfrac{17.9}{6.4} = 2.8\,\text{hr}$

0.8 hr = 0.8 × 60 min = 48 min   Answer: 2 hr 48 min.

Note on directions   Today we use a 360 degree system that always involves *three* figures; thus 5° becomes 005°. The topic of True and Magnetic bearings will be dealt with later in Chapter 4. Try to memorise these equivalents:

000°(or 360°) is **North**	180° = **South**
045° = NE	225° = SW
090° = **East**	270° = **West**
135° = SE	315° = NW

Arrange the following into two lists, one headed Direction and the other Distance. Latitude and longitude. Parallel rule. Dividers. Compass. Patent log.

Measuring a total distance involving several changes in direction; i.e. a circuitous or indirect route (see Diagram). What is the total distance around the island of Jersey using the Violet Channel?

Exercise 1   Checked from chart.

Exercise 2
2a   East Goodwin Light Vessel. *Note:* the exact position is at the centre of the vessel's waterline. This is the correct procedure for all such features.
2b   11.8 m wreck
2c   044° True   7.05 M
2d   243° True   7.73 M
2e   Check off answer below (2f).
2f   150° True   1.09 M
2g   12.8 M (This is best done with dividers set at 4 or 5 M increments.)

Exercise 3
3a   Southern Lighthouse of St Peter Port harbour.
3b   Barnouic Light Tower.
3c   210° True   30.5 M

Exercise 4
distance	*direction*
latitude	longitude
dividers	parallel rule
patent log	compass

Exercise 5   About 34 to 35 n. miles (M), depending on the exact route taken.

Charts and Publications

For the cruising yachtsman, especially 'going foreign', a good folio of up-to-date charts backed by a full range of supporting publications is essential. The charts should cover not only the intended cruising area, but harbours along the route which might be needed as a refuge in bad weather or because of a change of plan. The cost should be regarded as a good investment, ensuring as it does peace of mind.

Charts

Many yachtsmen, including myself, prefer Admiralty charts because of their wide range of cover, correction service and general high standard. Privately published charts are, however, used extensively because they fill several gaps in scale and coverage. Very often these charts include 'insets.' Insets are used to give extra detail of harbours and anchorages to a larger scale, thus avoiding the cost of additional charts. A good example is the Stanford chart included with this book.

Purchasing and inspecting an Admiralty chart

Before purchase check that your chart has been brought up to date by the chart agent. This can be verified by looking along the lower margin at the western (left-hand) edge for recent dates of 'small corrections'. (This form of correction is not found on privately published or Admiralty practice charts and so is not applicable to either of the charts referred to in this book.) These corrections are made by the main chart stockists and are taken from weekly Admiralty *Notices to Mariners*. More will be said regarding these Notices later in the chapter.

You now need to become thoroughly familiar with your new purchase. The obvious thing to check first is the units in which the soundings (or depths) are presented. This is shown under the title, and metric charts have the words DEPTHS IN METRES printed in the margin top and bottom. Metric charts are increasingly common, as fathom charts are being phased out. Next measure one or two distances and thus become familiar with the scale. Remember that on small or medium scale Mercator charts, the scale must be measured at the latitude of the area concerned.

Take a look at the compass rose and note the amount of variation, the date and the rate of change (the significance of magnetic variation is explained in Chapter 5). Familiarise yourself with any notes on the chart which may contain information important to you. These may be near the legend (title) or elsewhere. The most unusual example of the latter that I have come across, was on an Admiralty chart of the Morbihan in southern Brittany, and warned about the flight path of low-flying fire-fighting aircraft, taking up sea water between two of the islands! I remember that my younger daughter was most disappointed when they failed to put in an appearance.

A chart also gives important tidal information, which will be discussed in Chapter 3, particularly what is known as the *chart datum* — the horizontal level to which the soundings or depths of the sea bed are referred. On modern Admiralty charts this is approximately Lowest Astronomical Tide (LAT), the lowest predictable tide under average meteorological conditions. Otherwise an indication of chart datum can best be achieved by looking at the level of Mean Low Water Springs (MLWS) in the table giving tidal information. Here you will find listed various ports that appear on the chart and the appropriate low and high water heights for both spring and neap tides. By subtracting the height of Mean Low Water Springs (MLWS) from Mean High Water Springs (MHWS), find the spring range. If the MLWS level is more than 10% of this range, then the chart datum for that port and its immediate neighbourhood is probably at LAT (lowest astronomical tide). Such charts show absolute minimum depths and can be used without reservation. Where the MLWS level is less than 10% of spring range (frequently only 1–2%), then the datum is at a higher level. During abnormally large spring tides such charts must be used with caution, as sea levels can very occasionally fall below the datum line. This is of particular importance in the Channel Islands and other areas of large tidal range, such as the Bristol Channel. (Tides are discussed in more detail in Chapter 3.)

Foreign charts

Although these are expensive and difficult to obtain, it is well worth the effort should you be interested in a particular area. This implies that your need for a foreign chart will usually be for one of a large scale (greater detail). The latest official French charts are well produced and are easily understood, as most of the important symbols are similar to those used on Admiralty charts.

Arrangement of the chart folio

Charts are best arranged geographically rather than alphabetically. For this purpose a space for the Consecutive No. to be inserted is provided on Admiralty charts. It is also a good idea to keep a list of all charts on board, together with their number, scale, edition (date), and when purchased.

Publications directly supportive of the chart

A most useful booklet is an Admiralty publication known as NP 5011, *Symbols and Abbreviations used on Admiralty Charts*. In it is all you need to know about signs, symbols and abbreviations. Each double page gives the older fathoms convention on the left and the metric convention on the right; fortunately they are not dissimilar and many are identical. International standardisation means that many foreign charts use the same systems. This publication is not expensive and should be regarded as essential. It presents the information in a far more comprehensive and clearer way than any textbook or almanac, and in the correct colours.

Should you intend buying a number of charts, especially when planning a cruise, first obtain a copy of the Admiralty *Home Waters Catalogue* (NP 109), now issued free in sheet form. This will enable you to make a long and considered choice, rather then a hasty one in a busy shop.

Notices to Mariners

The weekly editions to cover a full year amount to a pile of texts over a foot in height. They are, however, very worthwhile for the serious cruising yachtsman. Fortunately they are available from Admiralty Chart Agents *Small Craft Editions* for home waters which are published five times a year. With their help you may correct not only your charts but also the other Admiralty publications you may possess, albeit after same lapse of time. Alternatively you may return your charts for correction by the agent, but this is costly and a chart will only be accepted if it is a recent edition. (The same applies to commercially printed charts, corrected by their publishers.)

Chart corrections

Because of the great deal of work that goes into creating them, charts are expensive. But a chart can soon become valueless — indeed dangerous — if it is not kept corrected. The only thorough way to keep charts up to date is by the *Weekly Edition of Notices to Mariners*, which can be posted to you every week by an Admiralty Chart Agent. These notices tell you what new editions of charts have been issued, and the corrections needed not only to charts but to volumes of the *Admiralty List of Lights* and *Admiralty List of Radio Signals* you may hold.

For an adequate update it is essential to use a good mapping pen, with waterproof purple ink. Symbols used for chart corrections should conform to NP 5011.

Yachtsmen are not likely to be interested in the details of oil wells or of soundings much above 10 metres. This reduces the task considerably. The principal corrections needed will concern navigational aids (lights, buoys, beacons etc) and changes to depths in harbour approaches.

Be sure to note 'small corrections' that you do in the space provided in the bottom margin of the chart on the left hand side.

Other publications

Here we have a distinction between publications for the professional mainer or globe-circling yachtsman and ones for those of us confined to home waters. The professional will tend to have on board specific books giving very detailed information world-wide, on such matters as lights, pilotage and radio navaids and communication. Yachtsmen will, for the most part, rely on a general book, more or less confined to home waters, called an *almanac*. The two most popular of these are *Reed's Nautical Almanac* and the *Macmillan & Silk Cut Nautical Almanac*. Both cover the British Isles and the adjacent Continent from Biscay in the south to the River Elbe in the north. Typically, almanacs will include the Collision Regulations or 'Rule of the Road', HM Customs regulations, coastal navigation, astro navigation, radio aids to navigators, communications, weather, safety, tides and harbour information.

Macmillan/Silk Cut is the more recent publication and much of its information is arranged regionally. *Reed's Almanac* has most information organised topically. The *Macmillan/Silk Cut* does have the advantages of being easier to use and of giving more pilotage information over a larger area, but *Reed's* tends to find favour with the traditionalist and some of it is set in clearer, larger type.

Pilot books

These expensive publications are of immense value to the cruising yachtsman, many of whom refer to them as 'Bibles'. To a great extent they replace the need for numerous large scale charts and at the same time give valuable local knowledge. They contain notes on pilotage, photographs, sketches, harbour plans and small 'chartlets'. No long cruise should be planned without one or more. Before purchase check the book's date of publication and its reputation for accuracy, as there are often alternatives and the depth and care of research and updating does, unfortunately, vary.

Tidal stream atlases

Although charts and almanacs give tidal stream data, it is much better and more convenient to have this vital information in atlas form. Passage planning is often started and subsequently dominated by references to the tidal stream atlas. There are several good alternatives to the Admiralty publication, notably the *Stanford Tidal Stream Atlas* and *The Yachtsman's Tidal Stream Atlas*. Both are written by the same author, Michael Reeve-Fowkes, and offer several advantages including quick-reference tidal heights, easy computation of tidal stream rates, and detailed coverage from Brest to the North Sea; I have found them to be accurate.

Manufacturers' manuals

All manufacturers' manuals should be kept on board, and here I refer particularly to those relating to navigational instruments. They can be of particular help should a fault develop while in a foreign port. The more information you have to hand the better in this situation. Radar and Decca based equipment produce a particular need for constant reference to handbooks in the early stages. In fact, those users of Decca who fail to do this are unlikely to learn how to get the best out of their sets.

Deck log

One other important document remains to be discussed — the deck log, in which several important entries need to be made at regular intervals. These entries are essential, for example when working up the boat's position by dead reckoning as described in Chapter 6. The form a log takes is immaterial; what matters is that it is filled in conscientiously. A ruled exercise book is perfectly satisfactory, and suggested headings are shown in the example in the Appendix. Regardless of the course ordered, it is important that the helmsman records the course that has actually been maintained. In order that this can be assessed accurately, entries should be made at least every hour, and more frequently in bad weather.

Summary of publications

Essential: *Symbols and Abbreviations used on Admiralty Charts* (NP 5011). Local tide tables. Manufacturers' manuals.

Essential for cruising: Almanac for current year, Pilot book.

Very useful: *Admiralty Notices to Mariners, Small Craft Edition*. Tidal stream atlas.

Useful: *Admiralty Home Waters Catalogue of Charts and other Publications* (NP 109)

Long distance cruising only: *Admiralty List of Lights, Admiralty Tide Tables, Admiralty List of Radio Signals*. Admiralty Pilots, various.

Exercises in the choice and use of publications

These questions, for the most part, may be taken at two levels. Should the reader be without an up-to-date almanac, then simply decide which type of reference book would best answer the question. Here questions have been chosen that can be answered by reference to a particular almanac, namely the 1987 edition of *Macmillan/Silk Cut*. In many cases a pilot book or reference to NP 5011, for example, would be both quicker and give a more complete answer. These questions have a strong French bias because most yachtsmen experience a greater need for information when sailing in foreign waters; for this reason a cruise around Brittany has been chosen as a theme.

Exercise 1

Which of the following publications would best answer these questions: (a) *NP 5011*, (b) *North Brittany Pilot*, (c) *Notices to Mariners*, (d) *Admiralty List of Lights*?

1 The meaning of a mysterious chart symbol?

2 Depths of water in the channel between Ile de Brehât and the French mainland?

3 Details of buoys marking a new wreck?

4 The exact characteristics of the Eddystone Light?

Exercise 2

You intend to cruise Brittany but need to keep in close contact with your family throughout the cruise. Make a list of VHF stations through which you could make link calls, together with the correct channel number. Work anticlockwise around the coast starting in the St Malo area and finishing at the Loire.

Exercise 3

Having rounded the SW tip of Brittany (Pte du Raz), you are approaching the Rade de Brest in poor visibility. (a) Which radio beacon would be most useful? (b) What is its frequency, call sign and range? (c) Is the beacon continuous or is the frequency shared?

Exercise 4

At what times, relative to high water, can you lock into the inner basin in the port of Lorient on the south coast of Brittany?

Exercise 5

You need to make a positive identification of the light exhibited from Le Grand Jardin rocks off St Malo. What are its characteristics?

Exercise 6

At what time should you leave St Malo in order to have a favourable tide for Paimpol? Answer to be given relative to high water (HW) Dover or HW St Helier.

Exercise 7

Is the tidal information on Brest given in GMT, BST or French Summer Time? *Note*: FST = GMT + 2 hr (time zone − 0200).

Exercise 8

On an Admiralty chart you find that a sounding has the letters M.fS.Sh. near it. What does this mean? Would it make good holding ground for an anchor?

Exercise 9

Is the River Treguier navigable at all states of tide?

Exercise 10

You are rounding Ushant for the first time and need good weather information. Apart from the British shipping forecast, what are the alternatives? Assume that your ability to translate a French language forecast is not adequate.

Exercise 11

What documentation should be carried on board a British yacht entering French waters?

Exercise 12

What is the VHF channel number for the port of Etel in southern Brittany through which pilotage directions may be obtained?

Chart questions

Exercise 13

Calculate the natural scale of the Stanford chart supplied with the text.

Exercise 14

Anse d'Yffiniac lies in the SW corner of the Stanford chart. What is the greatest drying height in the approaches to the port of Le Légué?

Exercise 1
1 a Admiralty publication *NP 5011*
2 b *North Brittany Pilot*, or a large scale chart
3 c *Notices to Mariners*
4 d *Admiralty List of Lights*

Exercise 2 Best source almanac or *Admiralty List of Radio Signals*, Vol. I Part 1. Note that only Coast Radio Stations (CRS) and not Port Radios (PR) provide link calls. (*Macmillan/Silk Cut* 6.3.12). St Malo, *Channels 01 and 02*. Paimpol, *Ch 84*. Plougasnou, *Ch 83*. Brest le Conquet, *Ch 26 and 28*. Lampaul (Ushant), *Ch 82*. Pont L'Abbé, *Ch 27*. Le Palais (Belle Ile), *Ch 25 and 87*. St Nazaire, *Ch 23 and 24*. Nantes St Herblain, *Ch 28*.

Exercise 3 Best sources are almanac or *Admiralty List of Radio Signals*, Vol II.
a Pte St Mathieu
b Frequency 289.6   Call sign **SM** . . . _ _   Range 20 M
c Not continuous; grouped with four other stations. It is no. 3 in the sequence. *Macmillan/Silk Cut* – Table 4(2) No. 611.

Exercise 4 Best sources: *North Biscay Pilot* or almanac. 1 hr either side of HW

Exercise 5 Best sources: Up-to-date large scale chart. *Admiralty List of Lights*. Almanac.
Fl(2) R 10s (two red flashes every 10 seconds), visible 15 M.

Exercise 6 Best sources: *Admiralty Tidal Stream Atlas* (Channel Islands). *Stanford Tidal Stream Atlas* (West or Central Channel). Almanac (*Macmillan/Silk Cut* 10.17.3)
Start 4 hr before HW Dover, or 1 hr after HW St Helier.

Exercise 7 Best sources: *Admiralty Tide Tables Vol. I* or almanac. All times are given BST, i.e. Time Zone – 1 hr (−0100).

Exercise 8 Best sources: *NP 5011* or almanac (*Macmillan/Silk Cut* 3.1.4). Mud, fine sand and shells. Good holding ground.

Exercise 9 Best sources: French large scale chart. *North Brittany Pilot*. Almanac. Appears to be navigable at all states of tide. The almanac is not, in this case, an ideal source of information.

Exercise 10 Best sources: Almanac or *Admiralty List of Radio Signals* Vol. VI. Ouessant Traffic broadcasts in English every 3 hr from 0150 GMT on VHF Channel 11, after an announcement on Channel 16. (*Macmillan/Silk Cut* Table 7(5)).

Exercise 11 Best source: Almanac.
Passports. Yacht's Certificate of Registration. HM Customs Form C1328. Boat must register under full registration procedure or be placed on the Small Ships Register (*Macmillan/Silk Cut* 2.5.3).

Exercise 12 Best source: *Admiralty List of Radio Signals*, Vol. VI, Part 1. Almanac.
Channel 13 (*Macmillan/Silk Cut* 10.15.27).

Exercise 13 At the bottom of the chart we have a 'Scale of Distance at Mean Latitude'. This measures 24.7 cm and represents 25 M. 1 M equals 1852 m (see Chapter 1), so
25 M = 1852 × 25 metres, or 1852 × 25 × 100 cm
$$\therefore \text{Scale} = \frac{1852 \times 25 \times 100}{24.7} \text{ or approx. } 1{:}187{,}450.$$

Exercise 14 18 ft. *Note:* although charted depths are in fathoms on this chart, drying heights are given in feet. On metric charts, depths and drying heights are both in metres.

Tidal Rise and Fall

Throughout the greater part of the British Isles and the adjacent coasts of Europe, a significant rise and fall of tide occurs and tends to dominate our sailing. In the Bristol Channel and around the Channel Islands the tidal ranges are spectacular, reaching 14 metres east of St Malo. In such areas it is possible to navigate in three dimensions. Rocks, and indeed whole beaches, 2 or 3 metres above water are sailed over with impunity a few hours later. This makes pilotage particularly exacting, as the scene alters so dramatically with each state of tide. Islands like the Ecrehou, the Minquiers and the Isles de Chausey are reduced in size by a factor of ten or twelve, as the spring tide sweeps in. This dramatic range of tide adds greatly to the enjoyment of such areas, but only if this movement is thoroughly understood and appreciated. Anchoring, for example, requires some thought in a river in which the depth may change by 12 metres in the time taken to do the shopping and have a meal.

The theory of tides

Tides (the vertical rise and fall of the sea level) are due to the gravitational effect of the moon, and to a lesser extent the sun. The attraction of the moon is not uniform over the surface of the earth: nearest the moon it is greatest, and on the far side of the earth it is least. This raises the sea level at places nearest to the moon and farthest from it, as explained in the diagram.

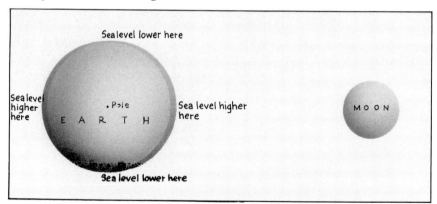

The combined effects of moon and sun are greatest when they are in a straight line with the earth, causing larger (spring) tides just after new moon and full moon, or about every 14 days.

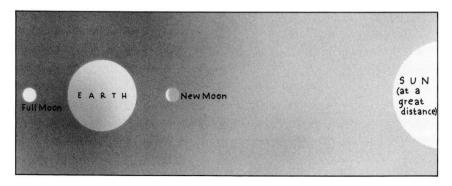

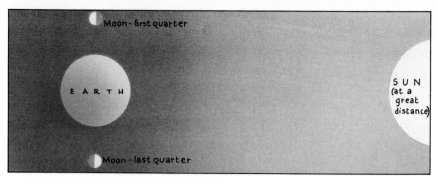

At the first and last quarters of the moon, the moon and sun form a right angle with earth, and their combined gravitational effect is least, resulting in smaller (neap) tides.

Because the earth rotates on its axis every 24 hours, the parts of the earth nearest to and farthest from the moon continually change, which is why the sea level at any one place rises and falls.

A lunar day is about 24 hr 50 min. In NW Europe two tidal cycles occur each lunar day, and these are called semi-diurnal tides with about 12 hr 25 min between successive high waters. Some parts of the world do not follow this pattern.

Tides are also affected by the surrounding land, for example in the Baltic and the Mediterranean. Along open coasts the tide usually rises at about the same rate as it falls on any given day, but in estuaries it usually rises more quickly.

The vertical rise and fall of sea level causes horizontal movement of the sea, called tidal streams, discussed in Chapter 4. These are quite distinct from ocean currents, which are caused by wind systems or by changes in the density of sea water through differences in temperature, or from river currents which affect tidal conditions in estuaries.

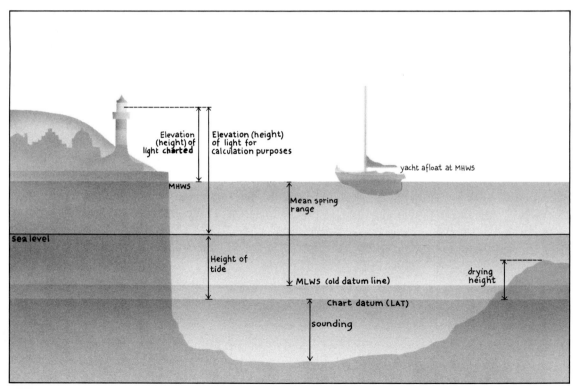

Tides: the basic reference lines

Study the accompanying diagram carefully. The figures, which are for St Peter Port, Guernsey, are less important than the levels they represent.

LAT   Lowest astronomical tide will be seen at the bottom of the diagram. It is the lowest height of tide predicted under average meteorological conditions. Very strong winds, storm surges or high barometric pressure may occasionally cause lower levels, however.

CD   Chart datum is the reference level for charted depths and drying heights on a chart. At any point the depth of water is the sum of the charted depth and the height of tide at the time. Modern practice is to establish chart datum at or very near to LAT, and this applies to metric charts of British waters. The same reference level is used for the predicted heights of tide in tide tables for British ports. Older (fathom) charts still in existence use a datum approximating to MLWS − slightly higher than LAT, as can be seen in the diagram. It is necessary to allow for this when using tide tables with most fathom charts, since otherwise the actual depth of water will be slightly less than calculated.

MLWS   Mean low water springs is the average height of low water at spring tides, and was used as chart datum on older (fathom) charts. It is quoted on Admiralty charts for selected ports.

MLWN   Mean low water neaps. Quoted on Admiralty charts for selected ports.

ML or MSL   Mean sea level (or half tide). This is a very important line, being the level about which every tide oscillates. Barring abnormal barometric pressures, this is the only level that is attained consistently by every tide. Roughly speaking, it can be thought of as occuring 3 hours either side of high or low water.

MHWN   Mean high water neaps. Quoted on Admiralty charts for selected ports.

MHWS   Mean high water springs. Quoted on Admiralty charts for selected ports. Also used as the datum line for the elevation (height) of shore objects such as lights, bridges, power cables etc.

HAT   Highest astronomical tide is the highest tide predicted under average meteorological conditions. It may occasionally be exceeded.

Range   Difference between consecutive low and high water heights.

Height of tide   Expressed in metres and always relative to chart datum. It can, very rarely, be negative, i.e. below datum.

Rise   Difference between the preceding low water height and sea level.

Exercise 1

Chart 5043

For these questions use the following tide table extract for Ramsgate.

Time	Height
0200	1.0 m
0800	4.0 m
1430	0.9 m
2030	4.4 m

A list of tidal levels for this port are given at the foot of chart/5043.

a By taking an average of the four levels quoted on the chart, find the MSL Ramsgate.

b At what times (a.m.) will MSL or half tide occur?

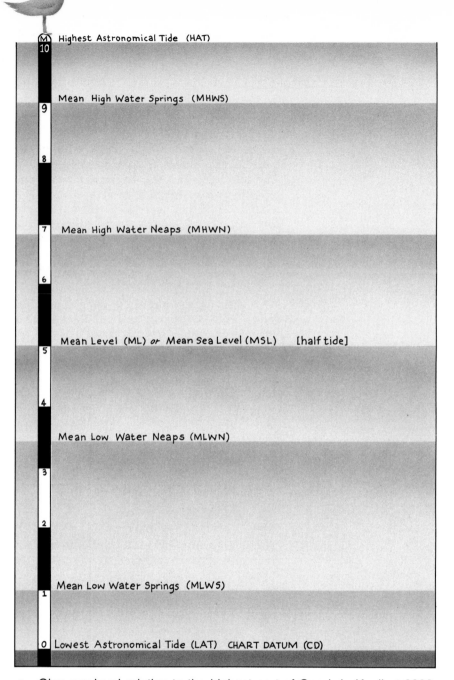

Highest Astronomical Tide (HAT)

10 M

Mean High Water Springs (MHWS)

9

8

7 Mean High Water Neaps (MHWN)

6

Mean Level (ML) or Mean Sea Level (MSL)   [half tide]

5

4

Mean Low Water Neaps (MLWN)

3

2

Mean Low Water Springs (MLWS)

1

0 Lowest Astronomical Tide (LAT)   CHART DATUM (CD)

c Give sea level relative to the highest part of Goodwin Knoll at 0200.

d Give sea level relative to the highest part of Goodwin Knoll at 0800.

e Give sea level relative to the highest part of Goodwin Knoll at 0500.

f Would the wreck on the Goodwin Knoll be visible at 1430?

g What is the height above sea level of the Fixed Green light at Ramsgate at 0500?

h What rise of tide has occurred between 0200 and 0500?

i What is the p.m. range?

Patterns, rhythms and tendencies

Some tidal patterns are general, others are peculiar to a locality. The fact that spring and neap tides alternate week by week is a general characteristic; whereas the fact that low water springs always occur in the early hours of the morning and again in early afternoon happens to be characteristic of the Channel Islands.

The lunar month

This is a span of time of just over 27 days in which the whole tidal cycle takes place. The cycle may commence with big spring tides followed by neaps, then the small spring tides, followed by the smallest of neaps. The latter have no official name, but could be called 'sub neaps'. They are particularly associated with the huge equinoctial springs that occur at the autumn and spring equinoxes. At St Peter Port, for example, 'sub neaps' have a range of only 2.0 m; a week later that range may be 9.5 m or more. Sub neaps produce their own difficulties, when the exceptionally small rise of tide produces insufficient water to navigate over marina sills and half-tide passages. Several marinas on the Brittany coast are unable to open over sub neaps and remain closed for a few days. Your enforced stay, if trapped inside, may or may not be welcome!

The weekly cycle

During any particular week the tides may be growing in size (making) or diminishing in size (taking off, or cutting). From one day to the next, the time of high water and that of low water advances by almost an hour. A useful point to remember is that springs advance by less than average, whereas neaps may advance by up to 2 hours.

The daily cycle

Each successive high water is, on average, just under 12½ hours apart, but the fall and rise in between is neither symmetric nor uniform, some ports having a more symmetrical curve than others. In most cases the rise and fall follows an approximate sine curve. The effect of this is to produce very little movement during the first and last hours, with half the total range being achieved in the middle hours. Many sailors find the 'Twelfths Rule' useful in approximating tidal rise and

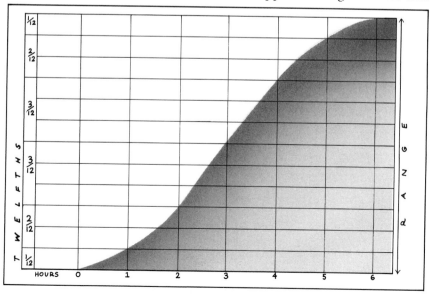

fall. It is very practical when anchoring, entering a drying harbour or sailing over a reef. Ideally a more accurate calculation should be made with pencil, paper and tables, but very often the time and opportunity are not available. To use the Twelfths Rule we must know the approximate times of either low or high water and the tide's range. The rule assumes that all tides rise and fall over a period of 6 hours, which as we have seen is not strictly true, especially on a falling neap tide when the period might well be 7 hours. The Twelfths Rule must therefore be regarded as an approximation, and in some places it can be misleading.

The Rule of Twelfths

Rise or fall of tide in 1st hr = $\frac{1}{12}$th of range

$$
\begin{aligned}
\text{2nd} \quad &= \tfrac{2}{12}\text{th} \\
\text{3rd} \quad &= \tfrac{3}{12}\text{th} \\
\text{4th} \quad &= \tfrac{3}{12}\text{th} \\
\text{5th} \quad &= \tfrac{2}{12}\text{th} \\
\text{6th} \quad &= \tfrac{1}{12}\text{th}
\end{aligned}
$$

Using the Twelfths Rule

Example 1 Find the height of tide at midday, when LW occured at 1000, height 2 m, range 6 m.

Rise in 2 hr = $\frac{1}{12} + \frac{2}{12} = \frac{1}{4}$ of 6 m = 1.5 m.

Height of tide at midday is therefore 3.5 m (2 m + 1.5 m)

Example 2 A yacht anchors in 8 m of water at 12.30. What depth will there be at LW? HW is at 1100 at 7 m, LW 1715 at 3 m.
Range is 4 m. Time is HW + 1½ hr. Therefore the tide will have fallen by approximately $\frac{2}{12}$ of the range or $\frac{2}{3}$ of a metre. Tidal range is 4 m, therefore $3\frac{1}{3}$ metres yet to fall. Approximate depth at LW will be $4\frac{2}{3}$ metres. *Note:* fall must be subtracted from echo-sounding, *not* added to LW height.

Exercise 2

a By how much will the tide rise in the 4 hr after LW if the range is 6 m?

b What will be the depth of water over a reef charted as 'drying 1.4 m'. Time is 2 hr before LW. Tidal range 4.4 m. LW height 2 m.

c A yacht anchors in 5 m of water at 1400. What depth will there be at LW? HW 0950 at 4.0 m, LW 1600 at 1 m.

d A yacht anchors in 7 m of water at 0845. What depth will there be at the next LW? HW 1140 at 9.4 m, LW 1805 at 0.8 m.

e A drying pad is situated 3.5 m above chart datum and being used by a yacht with a draft of 1.6 m. At what time will she ground on the pad? Use the following extract from the tide tables: HW 0915 at 8.5 m, LW height 1.5 m.

f For what length of time will the yacht be aground?

Accuracy of tidal predictions

Times and heights of tides can be predicted accurately provided that weather conditions are not abnormal. The classic example quoted each year in the *Admiralty Tide Tables* is the storm surge that occurred in 1953 in the North Sea. A low barometer and prolonged severe gales from the north raised levels by 2.7 m on the E coast of England and by even more on the Netherlands coast. Much damage and loss of life resulted.
A high barometer reverses the effect, depressing tides below their

predicted values. High pressure also delays the time of both low and high water. A divergence of 11 from the normal atmospheric pressure of 1013 mb will change the sea level by 0.1 m. This only applies if the abnormal air pressure is general and sustained, however. Wind direction and strength also affects tidal predictions, but each locality has its own characteristics and therefore no general guidance is possible.

Time zones

All the required tidal data for calculations will be found at the back of the book. Full information for St Helier in Jersey and for Dover is given. This data is quoted from the 1986 *Admiralty Tide Tables*. Both ports are Time Zone GMT: should you refer to foreign ports in the future, remember the time zone will be different. Tide times in French ports, for example, are given in the *Admiralty Tide Tables* (and almanacs) for 'Time Zone − 0100', which is the Standard Time kept in France. This is the amount (1 hr) which has to be deducted from the quoted times to produce Greenwich Mean Time. For example, 1300 in time zone −0100, is 1200 GMT.

Note that tide tables, like most navigational publications, take no account of British Summer Time (BST) or similar Daylight Saving Times (DST) in other countries. These must be allowed for during the months when they are in operation. BST is 1 hr ahead of GMT, and is the equivalent of Zone −0100. (For example, 0800 BST is 0700 GMT.) When cruising the western coastline of Europe in the summer months, three time zones will be encountered. Ashore, local time will probably be 2 hr ahead of GMT; most west European tide tables will be quoting GMT + 1 (or time zone −0100) and British tide tables will be using GMT − so beware! It is best to use one time only for all navigational purposes and keep GMT + 1 (or BST) on the chart table clock during the summer. This is particularly important for catching British weather forecasts. In a typical passage planning exercise, all three times will be encountered.

Increasing reference is being made in navigational publications to Coordinated Universal Time (UTC). Rest assured that for all practical navigational problems UTC is the same as GMT.

Tide tables and calculations − which system?

By 1987, both *Reed's* and *Macmillan/Silk Cut Almanacs* will be using the Admiralty system as printed in the 1986 tables quoted in this book. At last there will be a unified approach to this problem.

Using the Tide Tables

This is very straight forward, but some readers may wish to check the system. Exercises refer to tables in the back of this book.

Exercise 3

a Give the a.m. times (BST) and heights of LW and HW for Dover on May 24.

b What was the range that morning?

c By looking at the data to the right of the Dover graph, check if this tide was mean, spring or neap.

d Give the p.m. times (BST) and heights of LW and HW for St Helier, Jersey on May 20.

e What was the range?

f Was the tide a mean, neap or spring tide?

Finding sea level for any time

We have already seen how this can be done by the Twelfths Rule. This method is only approximate but can be worked out quickly without pencil and paper. A much more precise method is to use the tidal graph associated with each Standard Port.

Example to find the time at which sea level reaches 6 m after the p.m. LW on June 24, at St Helier.

First we need the relevant data from the St Helier tide tables:

LW St Helier at 1540 BST, 1.6 m
HW St Helier at 2115 BST, 10.9 m

Next, draw a line on the range graph joining 1.6 m to 10.9 m. We can now read off the time for *any* height, by transferring horizontally from the left hand graph to the right. Using the spring curve, the time for 6 m is 2 hr 50 min before HW, or 1825 BST. This ties in nicely with the Twelfths Rule because 6 m is approximately mean sea level (MSL) Jersey, and we would expect this to be at half-tide or HW minus 3 hr.

Exercise 4

a Using the data in the example, at what time would sea level have risen to 8 m?

b At what time *after* HW p.m. on June 24 will the tide have fallen to a height of (i) 7 m, and (ii) 2.5 m?

Exercise 5

a Your yacht has a draft of 1.5 m and is occupying a drying berth in Dover harbour. She is moored in the berth, which dries 3.0 m, at noon BST on Monday July 21. Calculate (i) the time at which she will ground, and (ii) the time at which she will float.

This type of problem must now be tackled in the opposite way: in other words the time is specified and the height to be calculated.

Example What will be the height of tide in St Helier at 1230 BST on June 19? LW St Helier at 1056 BST, 2.6 m, HW at 1652 BST, 9.9 m.

Next, find the time relative to HW, i.e in this case HW minus 4 hr 22 min. Draw the graph as before from 2.6 m to 9.9 m and find the height using a mean reading between neap and spring curves, since the tide is at approximately mean range on this date. Transfer horizontally as before from right hand graph (tidal curve) to left hand graph (tidal range). The answer in this example is 3.6 m.

Exercise 6

a What would be the height of tide, using the same data as in the example, at 1400 BST?

b Again, give sea level at 1515 BST.

Exercise 7

a Your ETA at St Helier inner marina is 1630 BST on Wednesday August 20. The height of the sill is 3.6 m. What is the depth of water over the sill on your arrival?

b The sill gate opens when the depth of water over the sill is 2.2 m, i.e. 5.8 m above datum. Find the time at which access to the marina should be possible, assuming a draft of less than 2.2 m.

Finding tidal heights is not difficult for Standard Ports but it is less easy for Secondary Ports. However, it may be of use to know that lists of tidal heights for a variety of English Channel ports, both Standard and Secondary, have been pre-calculated for each hour and all possible ranges by Michael Reeve-Fowkes and are available in the *Stanford Tidal Stream Atlas* and *The Yachtsman's Tidal Atlas*.

Anchoring calculations: will the boat be afloat at LW?

This calculation is similar to Exercises 4,5 and 6 but the 'fall' of tide has to be found by subtracting the LW height.

Example A boat seeks shelter from a NW wind by anchoring in the lee of the land W of Dover harbour on Monday, August 25 at 1650 BST. By how much can the tide be expected to fall? HW Dover is 1526 BST, height 6.4. LW is at 2255 BST, height 1.4 m. Using the Dover graphs as before, we find that the sea level height at 1650 BST or HW + 1 hr 24 min is 5.9 m. If the height at 1650 BST is 5.9 m and that at LW (2255 BST) is 1.4 m, then the tide will fall by the difference, i.e. 4.5 m.

Exercise 8

a Using the answer in the previous example, find the minimum depth on the yacht's echo-sounder that the skipper can accept, given that the draft is 1.5 m, 2.0 m is required under the keel, and that the sounder transducer is sited midway between keel and waterline. *Notes:* (1) with this knowledge, the yacht can be taken cautiously inshore until the minimum reading is achieved, thus gaining maximum shelter. (2) For convenience with tidal calculations it is easier if the echo sounder is one of the few that can be adjusted to read the actual depth of water.

b *Tidal fall calculation.* Find the drop in water level off St Helier when anchoring at 1900 BST on Thursday, July 31.

Finding times and heights for Secondary Ports

The process whereby data is transposed from the Standard to a Secondary Port is a little difficult because the differences are not constant. It is an area where a well appraised estimate is perfectly legitimate, but the mathematical approach must be given for those who prefer it. This process is essentially one of interpolation and is quite easy by a graphical method or with a simple calculator. Let us take the finding of the time of LW at Braye harbour on Alderney as the example. The standard port is St Helier, Jersey. The almanacs and the *Admiralty Tide Tables* (ATT) quote the following time differences for LW Alderney: + 25 min if LW St Helier is at or about 0200 or 1400 GMT. Plus 1 hr 5 min if LW Jersey is at, or about, 0900 or 2100 GMT. It would be helpful, at this stage, to study the appropriate data page at the back of the book, to see how the information is presented. Interpolation will be necessary when LW times at St Helier fall between the four times quoted.

First, two examples and an exercise where no interpolation is required.

Example find the time of LW Alderney, p.m. on Sunday, May 25.

LW St Helier	1357	GMT
Time difference for Alderney	+ 0025	
when LW St Helier is at 1400		
LW Alderney	1422	GMT (= 1522 BST)

Example find the height of HW Alderney, p.m. on Friday, August 1.

HW St Helier	8.1 m
Height difference for Alderney when HW	
St Helier is 8.1 m	− 3.4 m
HW Alderney	4.7 m

Exercise 9

a Find the time (BST) and height of HW at Folkestone (Standard Port Dover) p.m. on Saturday, May 10.

Interpolation of time and height differences

Usually when calculating times or heights of tides at Secondary Ports it is necessary to interpolate, in order to decide what time and height differences should be applied. Sometimes this can be done by eye, but often it is necessary to use either a graph or a mathematical calculation.

The graphical solution is straightforward, and is most useful when tidal calculations have to be repeated — say for a vessel's home port. Here are shown graphs for the LW time and height differences for Alderney (Braye) on St Helier. The construction is simple and can be done on any suitable squared paper to any convenient scale.

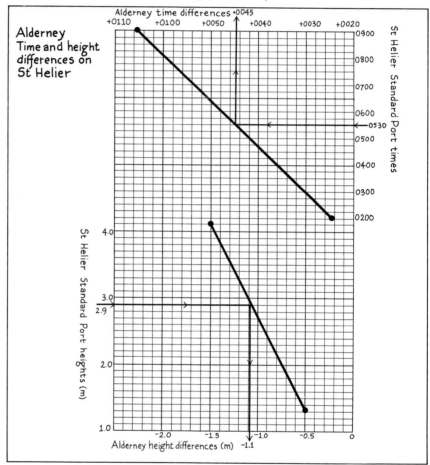

It is required to find the time (BST) and height of LW Alderney, when LW St Helier is 0530 GMT, 2.9 m. 0530 is between 0200 and 0900, at which times the time differences are +0025 and +0105 respectively. These two points are plotted (in the upper graph) and joined by a straight line. To determine the time difference for 0530, enter the graph at that point on the right hand side, proceed horizontally to the sloping line, and then vertically to the Time Difference scale at the top, and read off the answer: +0045.

LW St Helier	0530	GMT
Time difference for Alderney	+0045	
LW Alderney	0615	GMT (0715 BST)

With respect to the height of LW, another graph is drawn (the lower one shown) on which the Alderney height differences of −1.5 m and

−0.5 m are plotted against the St Helier LW heights of 4.1 m and 1.3 m respectively. These two points are joined. If the height of LW St Helier is 2.9 m, enter the graph at that point on the left and proceed horizontally to the sloping line, and thence down to the scale of Height Differences at the bottom. Read off the answer: −1.1 m.

This same example is now worked mathematically, to find the time of LW Alderney when LW St Helier is at 0530 GMT, 2.9 m. The reference times which matter in this case are 0200 and 0900. To these must be applied the time differences, +25 and +65 min. The following arrangement may be helpful:

+25 min   0200

              0530

+65 min   0900

We must now calculate three time differences:

Difference A = 40 min (65 − 25)
Difference B = 7 hr (0900 − 0200)
Difference C = 3.5 hr* (0530 − 0200)
We use the formula $\dfrac{C \times A}{B}$ or $\dfrac{3.5 \times 40}{7} = 20$

The time difference appropriate to 0530 is, therefore, 20 + 25 = 45 min. *Difference C must be expressed in hours: 0530 = 5.5 hr. This answer should seem sensible, in that 0530 is midway between 0200 and 0900, and 45 min is also midway between 25 and 65 min. Occasionally we will need to interpolate tidal heights and tidal stream rates in the same way. Usually we can make an acceptable estimate, especially if the situation is arithmetically simple.

Example finding the LW height, Braye Harbour, Alderney, if the St Helier height is 2.9 m. From the Differences table we see that when the tides are neaping (height St Helier 4.1 m), we subtract 1.5 m; but when the tides are springing (height St Helier 1.3 m), we subtract only 0.5 m. An estimation could run, 'A LW height of 2.9 m is roughly midway between 1.3 m and 4.1 m, therefore take a mean of 0.5 m and 1.5 m or 1.0 m.'

Interpolation
−1.5 m   4.1 m

              2.9 m

−0.5 m   1.3 m

Difference A = −1.0 m (1.5 m − 0.5 m)
Difference B =   2.8 m (4.1 m − 1.3 m)
Difference C =   1.6 m (2.9 m − 1.3 m)
$\dfrac{C \times A}{B} = \dfrac{1.6 \times (-1.0)}{2.8} = -0.57$ m
−0.57 m + −0.5 m = −1.07 m or −1.1 m

Note: because difference C is calculated by subtraction from the 1.3 m difference (−0.5 m), then the interpolation is related to −0.5 m. We could calculate C by referring to the 4.1 m difference (−1.5 m), in which case our answer would have to be deducted from −1.5 m. Always use a mentally calculated approximation to check interpolations carried out with the aid of a calculator.

The next example is more properly called an extrapolation. First let us consider the problem in context.

Example what is the height of HW Alderney, if the St Helier height is 12.0 m? *Note:* a 12 m tide at St Helier is not an impossibility: such exceptional tides occur frequently in all areas and are called equinoctial springs. They can exceed the mean HWS level by as much as 10%. Data:

−4.8 m   11.1 m

$\qquad\qquad$ 12.0 m

−3.4 m   8.1 m

Difference A = −1.4 m − (4.8 m − 3.4 m)
Difference B =   3.0 m   (11.1 m − 8.1 m)
Difference C =   3.9 m   (12 m − 8.1 m)

$$\frac{C \times A}{B} = \frac{3.9 \times (-1.4)}{3} = -1.8 \text{ m}$$

Because difference C was calculated using the 8.1 m (3.4 m), we must add our 'answer' (−1.8 m) to −3.4 to make −5.2 m. The amount by which to correct the St Helier height of 12.0 m is −5.2 m, or an Alderney height of 6.8 m. Extrapolations of this type also occur with 'sub neaps' and when calculating tidal stream rates.

Exercise 10

Find the time (BST) and height of LW and HW for St Peter Port, Guernsey, a.m. on Sunday, August 24.

Exercise 11

Find the time (BST) and height of LW and HW for Dungeness on Saturday, August 30 (p.m. only). Standard Port is Dover. 'Sub neap' tide.

A unique situation

Because the Standard Port, St Helier, is in the British time zone GMT, and the adjacent French Secondary Ports are in the Continental time zone, −0100, care must be exercised when calculating times of tides for these particular ports. As several of them are very popular with British yachtsmen, such as St Malo and Lezardrieux, the point is well worth making. Actually, the time differences include the time zone difference of 1 hr. Take St Malo as an example: the time difference for HW springs (0800 and 2000) is given as +0044. As this includes the allowance of 1 hr for time zones, the *real* difference is minus 16 minutes. The effect of this system is to create a situation in which the navigator changes zones without realising that he has done so.

Example time of HW St Helier is 2000 GMT, time difference for St Malo is +0044, therefore the time of HW at St Malo is 2044 **BST** (Zone − 0100). This rather strange system works perfectly well, once it is understood, but it does provide a trap for the unwary.

Exercise 12

a Find the time, BST, and height of LW and HW for Cancale, on Friday, August 22, p.m. only.

b What are the times of LW and HW at Cancale local time?

The exercises that follow will serve to reinforce the ideas already covered. The questions are in the form of practical situations and thus should also serve to put tidal problems in their correct context.

Exercise 13

a You wish to find the ship's position by a line of soundings near the Iles Chausey (standard port St Helier) at 1230 BST on Tuesday, August 26. Find by how much each sounding must be reduced in order to relate to chart datum. Full working for this question will be found with the answer.

b An hour and three-quarters later, at 1415 BST, you wish to enter the Sound at Chausey from the N. If the N passage has a bank drying at 4.8 m, have you sufficient water to pass over it with a draft of 1.8 m and a safety margin of 1.0 m?

c Assuming that the rising tide data is similar to that which you have on the graph, at what time relative to HW can you complete your pilotage into the Sound?

Exercise 14

The light at Dungeness is seen dipping and rising at 2315 BST on Monday, June 16. The light is charted as having a height of 37 m. MHWS Dungeness is 7.0 m. What is its actual height at 2315? Full working for this question is shown with the answer.

Exercise 15

a The Tréguier River (standard port St Helier) is shown as having a least depth of 1.2 m. Your ETA at the river estuary is 0215 BST on Friday, August 22. What depth is available at that time?

b If your draft is 2.0 m and you require 0.5 m under-keel clearance, at what time may you proceed upriver?
Full working for both questions is shown with the answer.

c If the anchorage in the Tréguier River is shown on the yacht's echo-sounder to be 5 m deep at noon BST, will she be afloat on the next LW? Use the graph already drawn, as the next LW height is higher by only 0.4 m, i.e. a height of 0.9 m.

Exercise 16

a The approach channel to Paimpol dries at half-tide (5.5 m). Assuming a draft of 1.5 m and a clearance under the keel of 0.5 m on a rising tide and 1.0 m on a falling one, find the hours at which you may enter either side of the p.m. HW on Sunday, May 11. *Note:* because the LW heights before and after HW are identical, only one graph line need be drawn.

Special tidal curves

It should be appreciated that in some places the rise and fall of the tide does not follow a regular or normal pattern. Between Swanage and Selsey on the South Coast of England individual tidal curves are shown in tide tables for various places. Because in this area low water is better defined than high water, these curves are drawn with their times relative to low water. In some cases a third 'critical' curve has been added for the range at Portsmouth at which the two high waters are equal at the place concerned. When using these curves interpolation should be between this 'critical' curve and either the spring or neap curve, as appropriate (Macmillan/Silk Cut 10.2.7).

Summary

Sources of tidal information
1 *Admiralty Tide Tables* Vols I, II and III give world coverage (Vol. I covers Europe).

2 *Reed's* and *Macmillan/Silk Cut Almanacs* cover the British Isles and adjacent European coasts.

3 Local tide tables, which as their name implies are specific to one locality. Two or three will prove very useful.

Tidal formulae
Rise = Range × Factor   (Factor is obtained from the centre of the tidal curve graph)
Factor = Rise ÷ Range   Sea level = Rise + LW height

Mean Sea Level (MSL) is strictly the mean of a large number of hourly heights of tide, over a considerable period. It is approximately the mean of MHWS, MHWN, MLWS and MLWN.

LW height $\simeq 2 \times$ MSL $-$ HW
Half range $=$ HW $-$ MSL

<table>
<tr><td>Answers to
Chapter 3</td></tr>
</table>

Exercise 1

a MSL Ramsgate 2.575 or 2.6 m.

b 0500 and 1115.

c Dry by 2 m. LW 1.0 m. Knoll dries 3 m.

d Covered by a depth of 1 m.

e Dry by 0.4 m.

f Yes. Dry by 0.9 m.

g Charted height 24 m (at MHWS). MHWS Ramsgate is 4.9 m. 0500 is half-tide of MSL. MSL $=$ 2.6 m. Thus 2.3 m has to be added. Height above sea level is 26.3 m.

h 1.6 m

i 3.5 m

Exercise 2

a $^9/_{12}$ths or 4.5 m

b 1.7 m

c 4.25 m

d 2.7 m

e Just before half-tide or about 1215. Height of tide at HW $+$ 3 is 5.0 m.

f About 6 hr.

Exercise 3

a LW Dover 0701 BST 0.7 m. HW Dover 1157 BST, 6.6 m.

b 5.9 m

c Spring (5.9 m is the mean spring range at Dover)

d HW St Helier 1636 BST, 9.5 m
   LW St Helier 2304 BST, 2.8 m

e 6.7 m

f Mean. Mean range is 6.9 (spring range 9.8 m, neap range 4.0 m).

Exercise 4

a   2 hr 10 min before HW or 1905 BST

b(i)   +2 hr 50 min, or 0005 on the 25th

b(ii)   +5 hr 25 min or 0240 on the 25th

Exercise 5

a Grounds at 1439 BST

b Floats at 2211 BST

Exercise 6

a 6.4 m

b 8.5 m

Exercise 7

a 0.4 m over sill, i.e. tidal height 4.0 m

b 1706 BST (3 hr before HW)

Exercise 8

a 7.25 m

b 1.3 m fall. Tidal height 5.5 m, time HW $+$ 4 hr 3 min.

Exercise 9

a 1240 BST, 6.6 m

Exercise 10

St Helier LW	0335 GMT,	1.6 m.	HW 0907 GMT,	10.3 m.
Differences	−0006	−0.3 m	+0012	−1.9
St Peter Port LW	0329 GMT,	1.3 m.	HW 0919 GMT,	8.4 m.

Answer: St Peter Port LW 0429 BST, 1.3 m.

HW 1019 BST, 8.4 m.

Exercise 11

LW Dungeness 1432 BST, 3.2 m
HW Dungeness 2052 BST, 5.5 m
Differences −19 min +0.2 m LW, −13 min +0.5 m HW.

Exercise 12

a LW St Helier 1447 GMT, 1.3 m. HW 2019 GMT, 11.1 m.

Differences	+0113	+0.7 m	+0049	+2.2 m
Cancale	1600 BST	2.0 m	2108 BST,	13.3 m

b 1700 FST 2208 FST

Exercise 13

a HW St Helier 1004 GMT, 9.3 m LW 3.1 m

Differences	+ 0046	+1.7 m	+0.7 m
Iles Chausey	1050 BST,	11.0 m	3.8 m

Using the St Helier graph with Chausey heights (11.0 m and 3.8 m) and a time *after* HW of 1 hr 40 min, the height of tide at Chausey at 1230 BST is 9.8 m. Soundings must be reduced by 9.8 m (about 5 fathoms).
b Anchor. Height of tide Chausey at 1415 is 7.3 m. Height required is 7.6 m (1.0 + 1.8 + 4.8 m).
c 2 hr 50 min before HW Chausey

Exercise 14

HW Dover 1730 GMT, 5.7 m. LW 1.6 m (on the 17th)
Differences −0015 +0.7 m +0.3 m
Dungeness 1715 GMT, 6.4 m 1.9 m
HW Dungeness 1815 BST
Time is therefore HW + 5 hr
Height of tide Dungeness is 2.8 m at 2315.
Additional height = 7 m less 2.8 = 4.2 m.
Light is 41.2 m high at 2315 BST.

Exercise 15

a St Helier LW 0233 GMT, 0.8 m. HW 0805 GMT, 11.0 m

Differences	−0017	−0.3 m	+0012	−1.4 m
Treguier	0216 BST,	0.5 m.	0817 BST	9.6 m

ETA is at LW. Therefore minimum depth in river is 0.5 m + charted depth 1.2 m = 1.7 m.
b Depth required is 2.5 m, therefore tide must rise by at least 0.8 m before entry is safe. Required height is 0.8 m plus the LW height of 0.5 m or 1.3 m. Using St Helier graphs, with Tréguier data, this height will be reached at HW − 4 hr 30 min or 0347 BST.
c No. Height of sea level at 1200 BST is 4.1 m, LW height = 0.9 m. Therefore 3.2 m to fall, leaving only 1.8 m (draft 2.0 m).

Exercise 16

a St Helier LW 2.2 m HW 1952 GMT 10.2 m

Differences	−0.8 m	+0038	−0.6 m
Paimpol	1.4 m	2030 BST	9.6 m

Height of tide required, rising = 7.5 m, falling 8.0 m
Using St Helier graph and Paimpol data, times are:
HW − 1 hr 50 min and HW + 1 hr 40 min, or 1840 BST to 2210 BST.

Tidal Streams

Because most of the British Isles enjoy a vigorous rise and fall of tide, they experience some of the most rapid and complex tidal streams. The Channel Islands and Goodwin Sands areas, covered by the two charts accompanying this book, are no exception; indeed, this was one of the main reasons that led to their selection. The two most crucial factors in navigating such waters are an accurate compass and good tidal stream information. Unfortunately the latter is not always as accurately tabulated as one would wish, though improvements are coming slowly all the time. While a healthy scepticism should be maintained with regard to the information supplied by tidal atlases, we must at the same time remember that this is the best information available and use it as intelligently as possible.

Sources of information

Some Admiralty charts have a certain amount of tidal stream information on them. The point to which the data refers is shown by a purple diamond. When they occur in roughly the right area, tidal diamonds can be used very effectively, but all too often they do not. The better alternative, especially for passage planning, is a tidal stream atlas, ideally the most recent one and to the largest scale. Commercial publications can often offer the yachtsman more than official sources. In the data Appendix at the back of the book will be found the essential parts of the *Stanford Tidal Stream Atlas* for the Western Channel. This deals extremely well with the area covered by the Stanford Channel Island chart. When working with BA chart 5043, the tidal diamonds will be used. In this way, both methods will be experienced.

Some tidal information is contained in Sailing Directions, for example the time that the stream turns at certain key places such as off harbours, headlands etc. Much can be learned about tidal streams by observing the actual flow round buoys, beacons, lobster pot floats and similar objects as you sail past. What you see may not always be what the tidal stream atlas would like you to believe.

Interpolation of rates

The speed at which the stream is running depends directly on the vertical movement of the tide. When using Admiralty tidal stream diamonds, two rates are given: the lower value is appropriate to a mean neap tide and the greater to a mean spring. Some method of interpolation therefore has to be used, and at the front of the *Admiralty Tidal Stream Atlas* will be found a graph by which this can be done. However, it is almost as quick to use the same interpolation procedure as explained in the previous chapter, when dealing with time and heights of tides at Secondary Ports.

Example   using chart BA 5043, we see that an hour after HW Dover the rates for diamond A are 3.0 knots and 1.6 knots, in the direction 046°T. We must now refer to the Dover tidal graphs, on which we see that a mean spring range Dover is 5.9 m and a mean neap range is 3.3 m. Let us suppose that the range for the day is 4.0 m.

Interpolation:

3.0 knots   5.9 m

                        4.0 m

1.6 knots   3.3 m

Difference A = 1.4 knots (3.0 − 1.6)

Difference B = 2.6 m (5.9 − 3.3)

Difference C = 0.7 m (4.0 − 3.3)

$$\frac{C \times A}{B} = \frac{0.7 \times 1.4}{2.6} = 0.38 \text{ or} \cong 0.4 \text{ knots}$$

Rate on a 4 m range Dover = 0.4 + 1.6 or 2.0 knots

If this seems a little complicated, then remember you can either use an estimation or solve the problem graphically. When we take a closer look at the *Stanford Tidal Stream Atlas*, we shall see that the rates are simply read off a table of precalculated results.

Exercise 1

a Given a spring rate of 3.2 knots and a neap rate of 1.8 knots, calculate the rate for a Dover range of 5.0 m.

b What is the rate if the Dover range is 6.2 m?

c What is the rate if the range (Dover) is 3.0 m?

Finding the correct hour

When using tidal diamonds and also tidal stream atlases, the correct choice of the line or page of data is of paramount importance. Rarely does one need to know the tidal stream for a specific moment in time: usually the need is to know the effect for a *period* of time, most often for one hour. For this reason it is always best to choose a page in the atlas (or a line from the tidal stream data on the chart), which is applicable for a time about half an hour *after* departure.

The half hour vector

To cover a tidal cycle of approximately 12½ hours, every atlas publishes 13 diagrams. It follows that one of the diagrams must be used for only 30 minutes. In most cases this will occur when the cruise you are planning spans the time between the last page (HW Dover + 6) and the first page (HW Dover − 6).

Example Finding the correct page. HW Dover is 1000 BST. Departure time is 1120 BST. The choice must lie between HW plus 1 hr (1100) or HW plus 2 hr (1200). At first, HW + 1 looks the better choice because it is nearer, but this would be unwise. The point is that information is required for the hour *commencing* at 1120. In other words, from 1120 to 1220, HW + 2 or 1200 is more relevant.

Exercise 2

a Choose the correct data from tidal diamond A on chart BA 5043, for a departure time of 1600 BST. HW 1930 BST, range 5.9 m.

b You are on passage NE from Elbow buoy (off North Foreland) to the East Margate buoy. Which tidal diamond should be used?

c What tidal stream data would be appropriate if the time at Elbow buoy was 0935 BST on August 5? Speed 5 knots.

d What would be the correct data had the time been 0130 BST and the date May 23?

Method for averaging tidal stream data

A problem which arises when using tidal diamonds is their location. Frequently it is necessary to take a mean of two or even three diamonds in order to assess more correctly the rate and direction of the stream for an area. If the directions are similar, then an averaging system is adequate. If, however, the relevant streams differ by more than 30°, the best solution would be to draw the vectors on the chart and use simple geometry to find the mean.

Using the Stanford Tidal Stream Atlas
Such an atlas is the only way to make satisfactory decisions about passage plans, because an overall picture is available of the stream directions and speeds hour by hour.

Exercise 3

a Use the atlas to find the best state of tide in which to leave Plymouth if bound E for Exmouth.

b If HW Dover is 0730 BST, what would be the best time to leave St Malo for Treguier?

c You wish to leave St Peter Port in Guernsey for Cherbourg. Winds are light and streams strong, so maximum help from the tide is needed. HW Dover is 0200 BST. What is the best time to sail?

Calculation of rates

Using the Stanford Atlas, no arithmetic need be done; it gives mean rates in knots, and the conversion table gives actual rates via Dover *Height*. This is more directly available from the tide tables than range and gives satisfactory results.

Exercise 4

a If the state of tide is 5 hr after HW Dover, what is the rate and direction of the tidal stream immediately S of Jersey? Dover height 5.5 m.

b Time is 0930 BST. HW Dover 1130 GMT, height 7.0 m. What is the maximum rate of the tidal stream in the Race of Alderney which lies between Alderney and the French mainland?

c What is the rate and direction of the tidal stream W of Brest 3 hr before HW Dover? HW Dover 5.3 m.

Exercise 5

Choosing the correct starting page
a HW Dover is 0943 GMT, height 6.9 m. On what state of tide (Dover) would you choose to start a cruise from Guernsey to St Malo at an average speed of 6 knots?

b You elect to sail from the SE point of Guernsey for St Malo at 0915 BST. Which page in the Stanford Atlas should be regarded as your starting page?

c List the rate and direction of the eight tidal streams you would encounter on your cruise to St Malo. Step across the atlas pages with a pair of dividers set at 6 knots, remembering to turn to a new page at each step. Write in the time on each of the eight pages before you start. The data you are gathering would form the basis on which you would shape your course to St Malo. Unfortunately, the atlas does not show the Minquiers reefs S of Jersey, so for the purpose of this exercise please ignore them. Remember to take a mean rate when between two sets of data.

Note: in this exercise there will be some latitude in the answers, due to differing methods of assessment.

d Extract the tidal stream data for a cruise from Chichester Harbour (E of the Isle of Wight) to Cherbourg at an average ground speed of 6 knots, and distance 72 miles. Starting time is 6 hr before HW Dover. Tidal height at Dover is 6.0 m.

e What is the approximate total set and drift of all these vectors?

f Was HW − 6 Dover the optimum time to leave Chichester?

Tide Races

It should be realised that the most dangerous sea conditions are often close to land − where the tidal stream runs strongly around rocks or headlands. An irregular sea bottom can cause considerable turbulence in such areas, so that when wind is against tide very dangerous overfalls can be created.

Tide races are shown on charts and are mentioned in sailing directions. They should be avoided when possible, but particularly in bad weather.

Summary and additional notes

1 Always plan your passage with the aid of the best tidal stream atlas available.

2 Always choose a page *later* than your departure time.

3 If sailing, choose tides which will help the yacht to windward, although this can make for rougher conditions.

4 If winds are strong and well established, allow for a wind-induced vector of up to 1 knot downwind. This will decrease windward streams and enhance those flowing downwind.

5 *Back eddies* In areas of strong tides there will often be useful back eddies inshore, if your pilotage is sufficiently good to make use of them. These eddies are especially reliable 1 to 2 hours before the tide is due to turn. In other words, the inshore stream often starts 1 to 2 hours early.

6 It is usually better to face a foul tide in the initial stages of a cruise than towards its end.

7 Avoid areas where tide races or overfalls are indicated on the chart.

Answers to Chapter 4

Exercise 1
a 2.7 knots
b 3.36 knots
c 1.64 knots

Exercise 2
a 224°, 3.2 knots. HW − 3 or 1630 BST.
b P
c 228°, 1 knot. (HW − 2. HW 1157 BST. Range 4.8 m.)
d 054°, 1.4 knots. (HW + 3, HW 2242 BST on 22nd. Spring range.)

Exercise 3
a 4 or 5 hr after HW Dover
b 3 or 4 hr before HW Dover, or about 0400 BST.
c About 0630 to 0700 BST.

Exercise 4
a 110° at 1.9 knots
b 9.8 knots (average of 9.5 and 10.1 knots)
c 180° at 1.9 knots

Exercise 5

a Either 1 hr before HW Dover, or HW Dover

b 1 hr before HW Dover, or 0943 BST

c

1st hr	0915 to 1015 BST	270° at 1.4 knots
2nd	1015 to 1115	260° at 1.6
3rd	1115 to 1215	240° at 2.7
4th	1215 to 1315	260° at 0.5
5th	1315 to 1415	135° at 1.8 (average)
6th	1415 to 1515	120° at 3.3
7th	1515 to 1615	090° at 3.1 (average)
8th	1615 to 1715	090° at 1.9

d

1st hour	HW −6	100° at 0.8 knots
2nd	HW −5	070° at 1.7 (average)
3rd	HW −4	085° at 2.6 (average)
4th	HW −3	085° at 2.6 (average)
5th	HW −2	085° at 2.0
6th	HW −1	085° at 1.2
7th	HW	SLACK
8th	HW +1	260° at 1.8 knots
9th	HW +2	270° at 2.8
10th	HW +3	270° at 3.0 (average)
11th	HW +4	270° at 2.1 (average)
12th	HW +5	SLACK

e East 1.2 M

f No. It would have been better to have had a net drift to the west, unless the wind was SE.

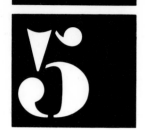

Chapter

5

Instruments for Coastal Navigation and Pilotage

Our present-day obsession with yacht instrumentation is a source of great joy to the chandler. Although the choice, installation, calibration and use of yachting instruments would form a volume in itself, the book would also require constant revision. We will, therefore, concentrate mainly on calibration and use. Choice is usually a matter of expense and is therefore a difficult topic on which to give advice. If an instrument is to be mounted outside in the cockpit, cheapness is rarely a virtue. The main advantage of expensive instruments lies not in their sophistication but in the robustness of construction and the degree of environmental protection offered. Certainly there is nothing like instrument malfunction to spoil a day's sailing: it is a source of constant irritation. The phrase 'buy in haste, repent at leisure' comes to mind, especially when ticking off the list of extras offered with a new boat. It is better to sail the first season with only basic but reliable instrumentation on board.

Installation

This requires a great amount of thought. Try one or two temporary installations first before cutting any irrevocable holes. Imagine the worst possible circumstances, in which the boat tries to emulate a rodeo horse combined with a submarine. Having fitted out my present boat in 1981, I soon learned the virtue of immobilising and sealing the sliding window above the chart area! White canvas covers for all cockpit instruments reduce temperature fluctuation on hot days, often causing the ingress of water as the temperature drops at night. The recent tendency to build instruments that are not controlled by knobs and switches but by pressure pads should make for easier installation, as long-term watertight integrity must be improved.

Steering and hand bearing compasses

Despite the amazing advances in the field of instrument technology, the mediaeval magnetic compass is still very much with us today. It is, therefore, imperative that we understand its inherent errors.

The angle of Variation

The first problem we must tackle is relatively straightforward. The error known as *variation* is due to the fact that the earth's geographical poles and their magnetic counterparts are not in the same location. Viewed from the British Isles, the Magnetic North pole lies approximately 6° to the west of True North. An accurate value can be determined from the east/west axis of any chart compass rose, which will also give the date on which the angle was measured and the annual rate of change. A typical reading taken from chart 5043 is: '6°15'W (1979), decreasing about 8' annually'. This implies that by 1987 this error will have dropped by some 64' or 1°04', and that variation should be regarded as being 5°11'W or approximately 5°W. We can, therefore, see that variation is aptly named, changing as it does with both its position on the earth's surface and with time.

Extraordinary variation Very rarely the normal variation reading for an area is distorted by a local magnetic field, probably due to a deposit of magnetic iron ore on the sea bed. Such an area is mentioned on the Stanford chart, SE of Roches Douvres at 49°06′N, 02°49′W as 'Magnetic Anomalies Reported'. In this area, by my own observations, variation increases abruptly from 6°W to 13°W.

Exercise 1

Calculate the variation for 1988 to the nearest degree, for the area SW of Guernsey, using the Stanford chart.

The fact that variation changes slowly with time, is a strong argument for the exclusive use of the outer, or True, compass rose, which is also much easier to read. Combined with the fact that information given in navigational publications is invariably expressed in True (not Magnetic) terms, the best policy is **Always Work True**. Therefore we need to convert magnetic bearings to true before plotting them, and to convert true courses to magnetic before steering them. A much quoted rhyme may be of some help: 'variation west magnetic best, variation east magnetic least'. By 'best' is meant greater. The following diagrams should make the reason for this clear.

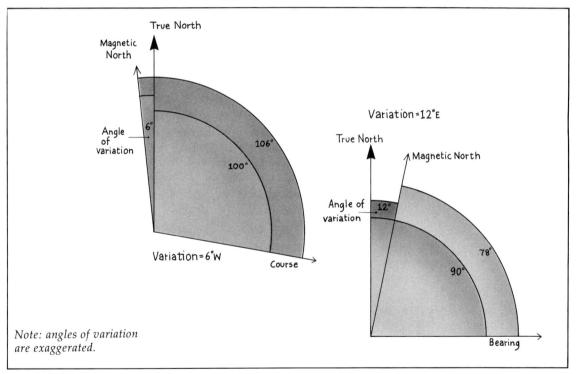

Note: angles of variation are exaggerated.

Exercise 2

a Convert 340° true to magnetic. Variation is 10°W.

b Convert 145° magnetic to true. Variation is 6°W.

c Convert 063° true to magnetic. Variation 4°E.

d Convert 180° magnetic to true. Variation 7°E.

e Convert 354° true to magnetic. Variation 8°W.

There are further exercises in compass conversion later in the chapter.

Deviation

This is a large and important topic and will be dealt with in depth. Deviation is one of the most common reasons for poor landfalls and indifferent fixes. It is due to unwanted magnetic fields on board the boat, deflecting or deviating the compass from its normal alignment. First let us understand some simple facts about magnetism. A common misconception is that 'metals are magnetic'. This is not true: only iron, cobalt and nickel have magnetic properties. Generally speaking only ferrous metals are detectably magnetic. This includes some, but not all, the stainless steels. The other source of unwanted magnetic field on board is electro-magnetism, mainly caused by conductors carrying large currents and electro-mechanical devices such as meters, motors and loudspeakers. Finally, it is worth noting that magnetism obeys an inverse square law: this means that if we are able to double the distance from a source of magnetism, the effect is reduced by a factor of four.

Causes of deviation   It should now be fairly obvious that the average boat has a host of potential sources of deviation in the shape of iron or steel ballast keels, engines, radios, wiper motors and many others. All we can do is to ensure that they are as far from the steering compass as is practicable. (Makers' installation instructions often mention a safe compass distance.) Be on the alert for mobile sources of trouble, such as screwdrivers (often strongly magnetised to assist in holding ferrous screws), steel cans, portable radios, loudhailers and the hand bearing compass itself. Some items are particularly bad as they will only deviate the compass when they are switched on. Check their effect on the compass when the opportunity arises while the boat is lying quietly, on at least two headings at right angles to one another.

Change of deviation error with heading   This is the bad news about this particular compass error. As the diagram should make clear, the alignment of Magnetic North, the source of deviation and the steering compass change as the boat is rotated. The phrase to bear in mind is **'deviation is heading dependent'**.

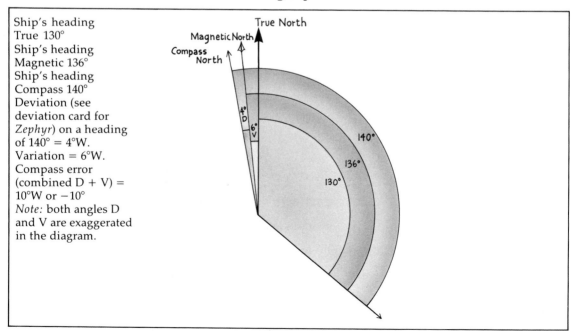

Ship's heading True 130°
Ship's heading Magnetic 136°
Ship's heading Compass 140°
Deviation (see deviation card for *Zephyr*) on a heading of 140° = 4°W.
Variation = 6°W.
Compass error (combined D + V) = 10°W or −10°
Note: both angles D and V are exaggerated in the diagram.

Deviation card for 'Zephyr'   *Engine off*   *Date made:*

Heading Compass	Deviation	Heading Magnetic	Heading Compass	Deviation	Heading Magnetic
000°	9°E	009°	190°	4°W	186°
010°	8°E	018°	200°	4°W	196°
020°	7°E	027°	210°	3°W	207°
030°	6°E	036°	220°	3°W	217°
040°	5°E	045°	230°	2°W	228°
050°	4°E	054°	240°	2°W	238°
060°	3°E	063°	250°	1°W	249°
070°	2°E	072°	260°	Zero	260°
080°	1°E	081°	270°	1°E	271°
090°	Zero	090°	280°	2°E	282°
100°	Zero	100°	290°	3°E	293°
110°	1°W	109°	300°	4°E	304°
120°	2°W	118°	310°	5°E	315°
130°	3°W	127°	320°	6°E	326°
140°	4°W	136°	330°	7°E	337°
150°	5°W	145°	340°	8°E	348°
160°	6°W	154°	350°	10°E	360°
170°	6°W	164°	360°	9°E	009°
180°	5°W	175°			

The main source of deviation is the engine. As this compass is installed directly in front of the engine, maximum errors will occur when heading east or west. On a north-south heading, magnetic north, the compass and the engine are in a straight line and no turning effect results.

A typical *deviation card* is shown next and it will be used in all future exercises when deviation value is required. How the data can be gathered, analysed and presented will be dealt with a little later on.

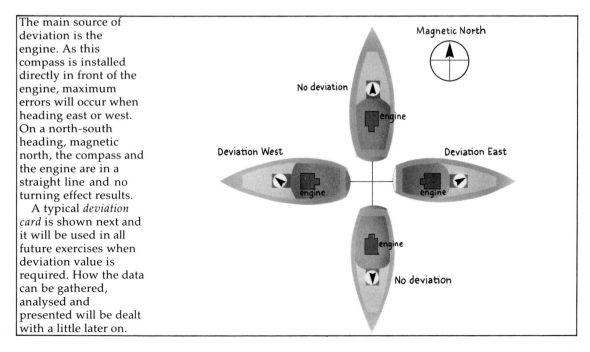

What is required, at this stage, are some rules that deal with both variation and deviation. A method developed during World War II for aircrew still makes most sense. There is a key word C ADE T and a phrase 'Cadbury's Dairy Milk Very Tasty'. CADET reminds you to **Add** **E**asterly errors when moving from **C**ompass to **T**rue. 'Cadbury's Dairy Milk Very Tasty' gives the key letters CDMVT for 'compass, deviation, magnetic, variation, true'. The following diagram shows how it sorts out the three reference lines or 'norths': compass north (where the compass is pointing), magnetic north (where the compass should be pointing), and true north (where we would like the compass to point and which is used for chartwork).

Example 1

Based on the previous diagram   What course (True) should be laid off on the chart if *Zephyr* is holding a course of 140° Compass? No leeway, variation 6°W. First write CDMVT; then write under each letter any known information. In this case we know C the compass course, D from *Zephyr's* deviation card, and V the variation. C ADE T reminds us to Add Easterly errors when moving from Compass to True. We are indeed moving from Compass to True, but because the errors are westerly they are subtracted:

C	D	M	V	T
140°	4°W	136°	6°W	130°

Check: the compass error (D + V) = −10°. 140−10° does equal 130°.

Exercise 3

Using the same deviation card for *Zephyr* and assuming variation to be 6°W, convert the following Compass courses to True. Check each one by first finding and then applying the compass error (D + V) to the Compass course.

a	300°C		**d**	345°C
b	170°C		**e**	033°C
c	195°C		**f**	352°C

In these exercises we have been converting from Compass to True, working from the steering position to the chart table. This does not happen very often, though one example is when beating to windward when the wind dictates the course steered. Another example is when taking bearings. On the majority of yachts, however, this is done with a hand bearing compass and such a compass can usually be used in an area free from deviation. As a result only variation has to be allowed for, following the 'variation west, magnetic best' rule. (How to discover a deviation-free area is dealt with later.) The more usual need is to convert from True to Compass, and this takes two small extra steps:

1   The CADET rule of Adding East, Compass to True has to be reversed because we are now converting in the opposite sense.

2   When using the deviation card, we must now enter it from the Magnetic heading column.

A worked example should make all clear:
A course of 330° True has been laid off on the chart. What course should be steered? Assume variation to be 6°W.

C	D	M	V	T
329°	7°E	336°	6°W	330°

Compass error +1°

Convert the following True courses to Compass courses, assuming variation as 6°W. Work all answers to the nearest degree.

a	048°T	**d**	009°T
b	255°T	**e**	358°T
c	313°T	**f**	002°T

Some further examples that also include allowance for leeway will be found towards the end of this section.

Determination of Deviation

Swinging the compass (not suitable for steel craft). A swing can be conducted in a number of ways but here it is best to concentrate solely on the easiest. The basic requirements in order to swing the steering compass by this method are a calm day, patience and a good hand-bearing compass.

If we are to find the errors in the steering compass by compass comparison, the reference compass must itself be in a position free from error. The most likely position is on the foredeck, although some go to the trouble of using a dinghy astern of the vessel, which needs more room and quieter waters than most harbours provide. On the foredeck, make sure that you are holding the hand bearing compass on the boat's centreline and sighting across the stemhead (bow). A quick comparison of the reading against a simultaneous reading of the steering compass will indicate whether there is a problem. If the compasses do agree, try the comparison again on two or three radically different headings. It is very unlikely that two such widely separated instruments could both be equally deviated: on the majority of yachts, a discrepancy will be found. The next move is to vindicate the position selected for the hand bearing compass. Use your local chart to find a convenient transit outside harbour but in sheltered water, a line formed by two charted objects, such as a church in line with a lighthouse. Take the boat, under power, at about 3 knots along that transit bearing. You should find that the line's magnetic bearing on the chart agrees closely with the reading on the hand bearing compass. Make sure you apply the correct amount of variation. Should they not agree a different onboard position will have to be found.

Now repeat the exercise by crossing the transit at right angles, in case the transit you have chosen is a heading on which deviation happens to be zero. (See *Zephyr's* deviation card on easterly or westerly headings.) The rest is simple but must be done carefully. Choose a calm area of water and steer 12 headings at 30° intervals, comparing compasses as you do so. It takes at least 12 seconds for a compass to settle down onto a new heading, so take your time. This is where the patience comes in, but remember (subject to occasional checks), that the swing you are conducting will be the basis of your navigation for the year.

Actions to be taken after swinging the steering compass

The first move, although in many cases it will be impractical, is to identify the cause of the deviation and remove it to a position more remote from the compass. If the errors have a strong bias in one direction only, check the fore-and-aft alignment of the steering compass. On the deviation card for *Zephyr*, for example, there is an easterly bias of 4°E: the compass would benefit from being rotated 2° to port. If the main errors occur on the N–S axis (as in *Zephyr's* case), the source of deviation lies to port or starboard of the compass. A fore-or-aft source of error produces maximum deviations when heading east or west. In most cases the 'cure approach' is impossible or only partially successful, and this leaves a clear choice between the following three courses of action.

1 Have the compass swung and adjusted by a professional compass adjuster.

2 Carefully tabulate the errors and always allow for them, as was done in Exercise 3 and 4.

3 Buy a remote reading compass system. This enables you to install the master unit in a deviation-free area and have the slave or repeater unit wherever is most convenient to the helmsman. This system, for a number of reasons, makes a lot of sense in motor yachts, especially fast power boats. It must, however, be backed up by a conventional compass in case of electrical failure.

One way or another, the compass should be checked by swinging at the start of every season. Adjustment may only be needed if alterations have been made to the boat, after considerable change of latitude (more than 15°), or if the corrector magnets start to lose some of their magnetism.

Steel-hulled craft obviously present special problems, and it is advisable for their compasses to be swung and adjusted by a professional compass adjuster, preferably annually.

When properly installed, remote reading compasses can be accurate and very convenient. Compass displays can be provided in different positions such over the chart table as well as in the cockpit. But such installations are very expensive and are normally only justified in larger yachts.

Heeling Error To test for this, note your compass reading when on an even keel under calm conditions and then read it again with the vessel heeled to her normal windward sailing angle: this should be done twice on different headings. How the necessary angle of heel

Mary Jane

April 1986 Engine running, radio on

Initial Deviations (data for graph)

Steering Compass	Hand Bearing Compass	Compass Heading	Deviation
000°	001°	000°	1°E
030°	033°	030°	3°E
060°	067°	060°	7°E
090°	101°	090°	11°E
120°	126°	120°	6°E
150°	152°	150°	2°E
180°	180°	180°	zero
210°	208°	210°	2°W
240°	236°	240°	4°W
270°	260°	270°	10°W
300°	294°	300°	6°W
330°	327°	330°	3°W
360°	001°	360°	1°E

combined with calm conditions is achieved may require some ingenuity, however.

Processing the results of a swing

This can be accomplished in four stages:

1   Calculate the errors.

2   Plot them as a graph.

3   Extrapolate a line between the plotted observations.

4   Tabulate for a useful number of headings.

The following example, from a hypothetical set of results, should be of some guidance.

Analysis of graph

Deviations are symmetrical, so the compass is correctly aligned. Major deviation on the E−W axis: the main cause of error is either forward or aft of the steering compass. Final deviation card (first quadrant only) as follows:

Mary Jane   April 1986 Engine running, radio on

Heading Compass	Deviation	Heading Magnetic
000°	1°E	001°
010°	2°E	012°
020°	2°E	022°
030°	3°E	033°
040°	4°E	044°
050°	6°E	056°
060°	7°E	067°
070°	8°E	078°
080°	10°E	090°
090°	11°E	101°

Alternative arrangement (first quadrant only)

Mary Jane   April 1986 Variation 6°W Engine running, radio on

Heading True	Heading Compass	Error
355°	000°	−5°
006°	010°	−4°
016°	020°	−4°
027°	030°	−3°
038°	040°	−2°
050°	050°	zero
061°	060°	+1°
072°	070°	+2°
084°	080°	+4°
095°	090°	+5°

This type of card recognises that conversions are usually made from true to compass. Also, variation remains constant unless the vessel

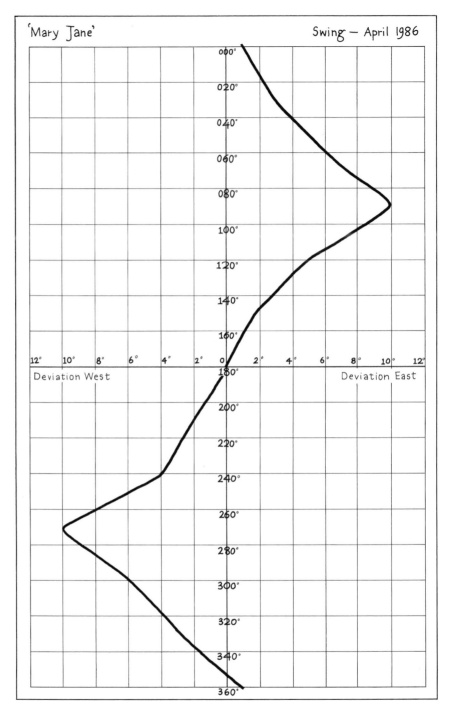

Swing — April 1986 'Mary Jane'

cruises over long distances. Magnetic readings are omitted as being of little use; they are merely a stepping stone between compass and true.

Having gone through this procedure, do not regard the matter as closed. Check to see if equipment that is sometimes 'on' and sometimes left 'off' creates a change in compass reading, for instance the 'transmit' mode of your radio apart from the 'listen' mode. If so care must be taken, especially in fog, to have them in the same mode as when the swing was conducted. If, on a sailing yacht, engine running alters the deviation, then two deviation cards must be carried.

During the ensuing season take an occasional check on the situation when the weather is calm, or if you suspect the compass accuracy. Remember in particular that any charted transit can be used to vindicate the position from which the hand bearing compass is habitually used, or to check the steering compass.

Exercises in the determination of deviation

Exercise 5

Use chart 5043

a Working off Deal in 51°13′.50N, 1°24′.00E, the pierhead light (2 FR) is observed in line with the conspicuous gasholders to the NW. The hand bearing compass gives a reading of 328°. Is this compass being used in a deviation-free area?

b The following data is the result of a swing conducted by compass comparison. Process it through, preferably via a graph, to the final dual-entry card. Only the first quadrant need be worked.

Steering Compass	Reference Compass
000°	356°
030°	023°
060°	056°
090°	089°

c Rewrite the card, incorporating variation of 5°W, so as to give direct conversion from a True heading to Compass and label accordingly.

Exercise 6

Allowing for leeway, deviation, variation (**leeway is covered on p 67**) These exercises are quite difficult and a rough diagram helps. The main dilemma, though only of minor importance, is whether to apply leeway first or last. The point to remember is that *deviation is heading dependent*. Use *Zephyr's* deviation card and assume variation to be 6°W.

a The yacht is closehauled on the starboard tack, steering a heading of 060°C and making 8° leeway. Find her true course. What is the approximate wind direction?

b The wind is SW. The yacht is heading 300°C. What is her water track True? Leeway is 6°. Which tack is she on?

c The navigator requires a course of 155°T. Wind is SW, leeway 10°. What heading should be steered by the helmsman? Which tack is the yacht on?

d The navigator calculates a course of 060°T. What is the heading Compass? The yacht is on the port tack, leeway 5°.

Exercise 7

Plotting bearings allowing for deviation
Assume bearings are taken off *Zephyr's* steering compass. Variation as per chart 5043. Remember: deviation is heading dependent.

a Give the latitude and longitude of the following position.
North Foreland Light 227°C
Foreness Point Coastguard 251°C
Elbow buoy 159°C
Zephyr's course was 142°C at the time the bearings were taken.

Compass installation: optimum criteria

1   The compass should be positioned directly in front of the helms-man.

2   Preferably in a head-up viewing position (difficult).

3   Protected. A non-magnetic stainless steel hoop is a good idea and makes a useful grab handle.

4   Deviation-free area of the boat (difficult).

The patent log

The yacht's log is a basic necessity and is complementary to the compass. It provides essential distance measurements and usually a speed indication, both relative to the water. Which type to buy is outside the scope of this text and is best discussed with your chandler. The most common choice is now a hull-mounted impeller or paddle-wheel, with a distance travelled readout at the chart table and the water speed indicator in the steering area. But the merits and simplicity of a towed log, where the rotator is well removed from the boat's hull, should not be overlooked.

Calibrating the log   The basic requirement is, once again, a calm day and to this should be added a stopwatch, pocket calculator and a crew of two or three. Two methods may be used. The first is by far the simpler and will more than suffice, if an area free from tidal streams can be found for the duration of the calibration. In fact, negligible errors will be incurred provided the tidal stream rate is less than 10% of the yacht's speed. With a pocket calculator a measured mile is not needed: simply choose two points between which you can safely travel a straight, accurately measured distance. Buoys should not be used as their charted positions are not sufficiently accurate. Use the largest scale chart available, making sure that your proposed route is parallel to the general run of the tide.

Procedure   On passing the starting point, under power at normal cruising speed and in a straight line, note the time and the reading of the log. Record its water speed reading every minute. As you approach the far mark, do not slow down or alter course: simply note the time and log reading as you pass. Now run the check in the opposite direction, cancelling out any weak tidal stream effect. *Note:* a 10% difference in elapsed times indicates that there was too much tide, in which case use method 2 or try again at slack water.

Example: breakwater light in line with church spire to Falls Rock beacon post, 1.89 M.

Time for run 1 = 20 min 40 sec
Time for run 2 = 19 min 10 sec

Distance logged on run 1 = 2.01 M
Distance logged on run 2 = 1.94 M

Water speed reading (average of 10) = 6.1 knots

Total distance covered = 3.78 M (2 × 1.89)
Total distance logged = 3.95 M (2.01 + 1.94)

Error = 0.17 M in 3.78 M or 4½%
Log over-reads by 4½%

Water speed indicator check
Total time taken to run 3.78 M was 39 min 50 sec, or 39.83 min.

$$\text{Speed in knots} = \frac{\text{distance in n. miles}}{\text{time in hours}}$$

$$\text{Speed} = \frac{3.78}{39.83} \times 60 \text{ knots}$$

Speed = 5.7 knots. Error is 0.4 knots or 7% over-reading.

Most logs can be adjusted and the manufacturer's instructions should give full details. A re-run is desirable as a double check after adjustment. In some models, distance and speed can malfunction separately, so both should be checked occasionally.

Exercise 8

Distance run A to B = 1.68 M

Time for run A to B = 17 min 5 sec

Logged distance A to B = 1.59 M

Time for run B to A = 16 min 40 secs

Logged distance B to A = 1.64 M

Average observed speed = 5.7 knots

a Find the percentage log error.

b Find the percentage water speed error.

We will now move onto method 2. This is more complex but has the advantage of not requiring slack water, although it is still advisable to choose an area of *weak* tidal streams. The following admittedly extreme example will illustrate the usefulness of method 2.

Example Distance from A to B is 1.00 M, boat speed 6.0 knots. Tidal stream 3.0 knots. This means that the vessel will have a ground speed of 9 knots in one direction and only 3 knots coming back. Taking account of log readings only, we get a completely distorted view of the log's accuracy.

Time A to B is one-ninth of an hour or 6⅔ minutes, during which time an *accurate* log could be expected to record 0.67 M (remember, 6 knots is 1 cable or 0.1 M per minute). Coming back, the ground speed is only 3 knots but the water speed indication will still be 6 knots. Because logs record distance through the water, it will read 2.0 M (20 min at 6 knots). We therefore have a colossal apparent error.

Distance covered over ground = 2.0 M

Total distance logged = 2.67 M

Error is 0.67 M in 2.0 M, or an apparent over-read of 33.5%.

Let us see how method 2 would cope with this situation. It is based on speed measurements, rather than distance, and will appeal to the meticulous. The main formula is

$$P = \left(\frac{(G_1 + G_2)}{(W_1 + W_2)} - 1 \right) \times 100$$

where P = percentage error, G_1 and G_2 are ground speeds for runs 1 and 2, and W_1 and W_2 are water speeds averaged from observations on runs 1 and 2.

Example To find the percentage error of the water speed indicator:

Distance run A to B = 1.0 M

Time A to B = 6⅔ min or 9 knots = G_1

Time B to A = 20 min or 3 knots = G_2

Average water speed (indicated) A to B = 6 knots = W_1

Average water speed (indicated) B to A = 6 knots = W_2

$$P = \left(\frac{(G_1 + G_2)}{(W_1 + W_2)} - 1 \right) \times 100$$

$$P = \left(\frac{(9 + 3)}{(6 + 6)} - 1 \right) \times 100$$

$$P = (1 - 1) \times 100$$

$$P = 0\%, \text{ i.e. water speed error is zero.}$$

Again, we do not need a measured mile when using method 2 if using a basic calculator. No observations are made from the distance recorder of the log. This is an advantage if the log reads only to one-tenth of a mile. It does mean that a subsequent check has to be made in which the accurately calibrated water speed indicator is used to check the distance meter.

Example Calibrating a water speed indicator by method 2.

Distance from A to B = 1.46 M

Time from A to B = 14 min 11 sec

Time from B to A = 17 min 27 sec

Average of 14 water speed readings A to B = 6.14 knots

Average of 17 water speed readings B to A = 6.05 knots

$$\text{Speed} = \frac{\text{distance (M)} \times 60}{\text{time in minutes}}$$

$$G_1 = 6.18 \text{ knots } \left(\frac{1.46 \times 60}{14.18} \right)$$

$$G_2 = 5.02 \text{ knots } \left(\frac{1.46 \times 60}{17.45} \right)$$

$$W_1 = 6.14 \text{ knots} \qquad W_2 = 6.05$$

$$P = \left(\frac{(G_1 + G_2)}{(W_1 + W_2)} - 1 \right) \times 100 \qquad \therefore P = \left(\frac{11.2}{12.19} - 1 \right) \times 100$$

$$P = (0.919 - 1)\ 100 = -0.081 \times 100 = -8.1\%$$

Water speed meter over-reading by 8%.

Exercise 9

a Find the percentage error of the water speed indicator from the following data:

Distance run A to B = 2.05 M

Time A to B = 22 min 15 sec

Time B to A = 17 min 20 sec

Average water speed readings A to B = 6.42 knots

Average water speed readings B to A = 6.56 knots

b An accurately adjusted water speed indicator is kept at a constant 5.0 knots reading, for 45 minutes exactly. The initial log reading was 12.4 M, and the final log was 16.2 M. What is the percentage error in the distance meter?

Remember: distance (M) = Speed (knots) × Time (hours)

In this exercise we will use higher speeds and assume that the water speed indicator is either fluctuating too much to be read accurately, or is non-existent. The speed through the water is, therefore, calculated from log readings.

a Find the log error from the following data:

Distance of run A to B = 1.32 M

Time A to B = 4 min 18 sec

Time B to A = 4 min 54 sec

Log readings:

Initial at A = 12.12 M
at B = 13.54 M

Initial at P = 13.72 M
at A = 15.21 M

We have dealt thoroughly with log and compass because they are of fundamental importance. The remaining navigational instruments do not require such exhaustive research.

The echo-sounder

Modern echo-sounders are cheap, accurate and reliable. They are the one electronic item which certainly deserves a place in any cruising yacht. They come in various types and at various prices, but they all work on the same principle. An electronic pulse is transmitted downwards from a transducer in the bottom of the boat, and the time that it takes to be received back is amplified and converted into some form of readable display.

Cheaper sets usually display the depth by means of a rotating arm, on which there is a flashing neon light or light emitting diode. A flash appears at the zero on the scale, which shows that a pulse has been transmitted, and the echo from the sea bed produces a second flash. The speed of sound in water is constant for all practical purposes, and since the arm rotates at a steady speed the time for the pulse to go to the bottom and back (and hence the depth) is shown by the angular displacement between the two flashes.

The receiver will also record echoes from other objects, such as fish or seaweed, between the transducer and the sea bed. With experience the type of bottom can be distinguished.

Other forms of display consist of a meter or a digital readout, but the most sophisticated sounders have a visual record (where the returning echo makes a mark on a moving sheet of paper, providing a permanent record of echoes form the sea bed and the water above it) or a cathode ray tube. These two are mostly used by commercial fishermen.

Many echo-sounders are fitted with a shallow water alarm, which is useful when cruising short-handed.

It is a simple matter to check an echo-sounder against the lead line, and this should certainly be done when the sounder is installed and at intervals subsequently. For convenience in use, the instrument should be adjusted to show the *actual depth* of water (not the depth below the transducer, or the depth below the keel).

Despite the merits of an echo-sounder, many experienced yachtsmen still carry an old-fashioned lead line in reserve. It can be useful to see whether the anchor is dragging, should no shore objects be identifiable, and for sounding from the dinghy.

Lead lines can still be purchased, marked in the traditional way or can be made up according to one's own system. 'Armed' with tallow they are also useful in assessing the bottom consistency i.e. sand, mud etc.

The radio direction finder (RDF or DF)

These instruments have become very sophisticated in recent years and can cost, in automatic form, almost as much as a Decca set. This, I feel, must spell their eventual demise in all but their cheapest versions.

The RDF set attempts to sense the bearing of an incoming radio signal by using an aerial with directional properties. These signals emanate from non-directional radio transmitters (beacons) installed usually in major lighthouses and airports. They are identified by their unique Morse code call signs, their frequency and their time of broadcast on shared frequencies, details of which are given in almanacs.

Most RDF sets in yachts are fitted with ferrite rod aerials, giving a minimum signal when pointed in line with the bearing of the station, and incorporating a small compass. The operator tunes to the beacon frequency and listens for the Morse identification signal (ident), after which a marine beacon transmits a continuous dash lasting about 22 seconds. This is the opportunity for the operator to swing the aerial and its compass so as to determine the point of minimum signal (or null) in his earphones. Some sets have a visual signal strength meter to assist this. The bearing obtained from the compass can then be plotted on the chart, in just the same way as a visual compass bearing of an object. Some aircraft beacons are suitable for marine use, and these normally transmit their Morse ident continuously.

It should be noted that sometimes (for example where taking bearings of a beacon in a light vessel or an offshore lighthouse) it is possible to plot the reciprocal of the actual bearing, unknowingly, so care is needed.

In some cases the receiver is fixed and the aerial rotates on it against a graduated 360 degree scale, which will then give the bearing of a beacon relative to the vessel's heading. With this type of set it is necessary to read the ship's compass at the same moment that the operator identifies the null and takes its bearing. This requires good coordination, and can be difficult.

RDF works on the assumption that the incoming signals travel in straight lines, but this is often not the case, as explained below. The key to success is constant practice and careful calibration.

Sources of error

1 *Quadrantal error* is due to the reflecting properties of the yacht's superstructure and rigging. The effect is to produce a series of errors in the yacht's environment very akin to deviation in a compass.The best answer is to conduct a RDF swing on a visible beacon and list the resulting quadrantal errors. Some sets have quadrantal error correction facility.

2 *Dawn and dusk errors*. Bearings taken at these times are not to be relied upon for about half an hour either side of sunrise or sunset.

3 *Refraction and reflection*. Radio signals travelling across land or leaving a coastline obliquely may be bent (refracted). At night incoming signals may have come via the ionosphere, high in the earth's outer atmosphere, from which they have been reflected. Both effects create errors which cannot be detected by the user.

One might ask, at this stage, why anyone bothers. Until recently, better alternatives have been too costly, too bulky and too heavy a drain on batteries. When out of sight of land RDF was the only system on offer by which a position could be measured.

In defence of RDF one might say that:

1 It works well when used as a homing device, i.e. when your desired track lies directly towards a beacon. This, of course, is its aeronautical use.

2 It is an essential backup to Decca should Decca fail in any way. For this reason, your back-up RDF set must be powered off its own internal batteries.

3 Used purely as a radio receiver, the RDF set can often provide valuable weather forecast information.

Calibration: see remarks on quadrantal error.

Practice is important. Try taking a bearings when the yacht's position is accurately known by other means, and in different sea conditions.

Preparation: To make the incoming signals as clear as possible use the nearest beacon(s) available, and do all you can to suppress radio interference or 'noise'.

1 Suppress the ignition system of petrol engines.

2 Fit switches to the field coil of alternators.

3 Fit 'Duff' brushes to the propeller shaft (see chapter 10).

VHF direction finding (emergency use)

Several VHF DF stations are located around the British coast, and some along the French coast. They are intended *for emergency use* when, on request from a vessel in difficulty, the station transmits the bearing of the vessel from the DF site.

The initial call should be on Ch 16. Once communication is established the yacht should transmit on Ch 16 (for distress situations only) or on Ch 67 for lesser emergencies so that the station can obtain the bearing.

Details of stations are given in almanacs (*Macmillan/Silk Cut* 4.3.2). Note that in the case of Jersey, Ch 82 is used instead of Ch 67.

Radar

Expensive, bulky and power hungry, radar is, nonetheless steadily gaining in popularity, especially aboard the larger yacht. As a means of avoiding collision in bad visibility it is without peer. Radar can be used to fix position via a bearing and range, provided that the target can be correctly identified. However, it does require a great deal of practice, both to tune the set and to interpret the information displayed. A radar set in the hands of a novice operator is, in fact, a source of danger. Up to now the cheap end of the radar market has proved to be unsatisfactory. At this moment radar requires a big commitment in both time and money, especially time.

Decca

This and other electronic systems are covered in Chapter 10.

Summary

Instrument	Function and importance	
Compass	Directional	Essential
Hand-bearing compass	Position fixing and collision avoidance	Essential for cruising
Log	Distance and speed	Essential for cruising
Echo-sounder	Depth	Essential
Radio direction finder	Position fixing	Essential for cruising
Decca	Position fixing and navigational computer	Should be backed up by inexpensive RDF. Will soon be regarded as essential for cruising.
Satellite navigator	Position fixing	System of the future. At present, excellent for ocean voyaging.
Radar	Collision avoidance and position fixing	Extremely useful in shipping lanes and other congested areas

Answers to Chapter 5

Exercise 1

6°W to the nearest degree.

Variation 6°40'W (1983) decreasing about 9' annually.

5 years at 9' = 45'   6°40' less 45' = 5°55'

Variation 1988 is 5°55'W, or 6°W to the nearest degree.

Exercise 2

a   350°M

b   139°T

c   059°M

d   187°M

e   002°M

Exercise 3

	C	D	M	V	T	Compass error
a	300°	4°E	304°	6°W	298°	−2°
b	170°	6°W	164°	6°W	158°	−12°
c	195°	4°W	191°	6°W	185°	−10°
d	345°	9°E	354°	6°W	348°	+3°
e	033°	6°E	039°	6°W	033°	0
f	352°	10°E	002°	6°W	356°	+4°

Exercise 4

	C	D	M	V	T	Compass error
a	050°	4°E	054°	6°W	048°	−2°
b	261°	zero	261°	6°W	255°	−6°
c	314°	5°E	319°	6°W	313°	−1°
d	007°	8°E	015°	6°W	009°	+2°
e	354°	10°E	004°	6°W	358°	+4°
f	359°	9°E	008°	6°W	002°	+3°

Exercise 5

a Yes. Transit is 323°T plus variation of 5°W gives 328°M.

b

Ship's head Compass	Deviation	Ship's head Magnetic
000°	4°W	356°
010°	5°W	005°
020°	6°W	014°
030°	7°W	023°
040°	6°W	034°
050°	5°W	045°
060°	4°W	056°
070°	3°W	067°
080°	2°W	078°
090°	1°W	089°

c

Ship's head True	Ship's head Compass	Error
351°	000°	−9°
000°	010°	−10°
009°	020°	−11°
018°	030°	−12°
029°	040°	−11°
040°	050°	−10°
051°	060°	−9°
062°	070°	−8°
073°	080°	−7°
084°	090°	−6°

Exercise 6

a 049°T (Apply deviation first: 3°E) Wind about ESE, say 100°
b 304°T (Apply deviation first: 4°E). Port tack
c 176°C (Apply leeway first in order to find the heading. Deviation is 5°W.)
d 058°C (Apply leeway first. Deviation 3°E.)

Exercise 7

a 51°25′00N, 1°30′00E Perfect fix (no cocked hat) deviation for all bearings 4°W. Compass error −9°.

Exercise 8

a Just under +4% (log is under-reading), i.e. log readings must be increased by 4%.
b Calculated speed is 5.97 knots, observed speed is 5.7 knots: ∴ percentage error = 4.5%. Indicator under reads by 4½%

Exercise 9

a −2.73% Log over-reads by 2¾%
b −1.3% *Conclusion:* Log now has negligible error on distance readout. 1% is probably inside the experimental error.
Note: a slight gain in accuracy would be achieved if the timing was taken from when the last figure had just appeared on the log, to a similar change say 2 or 3 miles farther on. This is of particular benefit for a log reading to only one decimal place.

Exercise 10

a Error −9% Log over-reads, i.e. all log readings must be reduced by 9%.

Chapter

Working up Ded Reckoning Positions (DR) and Estimated Positions (EP)

Definitions

Before discussing the various techniques of determining positions on the chart it is as well to clarify the definitions of certain terms which have been adopted recently and espoused by the RYA.

Track the path followed or to be followed from one position to another. This path may be over the ground (*ground track*) or through the water (*water track*).

Track angle* the direction of a track.

Track made good the mean ground track actually achieved over a given period.

Heading the horizontal direction of the ship's head at any moment. (This term does not necessarily require movement of the vessel.)

Course (Co) the intended heading.

Course to steer the course related to the compass used by the helmsman.

Set the direction towards which a current and/or tidal stream flows.

Drift the distance covered in a given time due solely to the movement of a current and/or tidal stream.

Drift angle* the angular difference between the ground track and the water track.

Leeway the effect of wind moving a vessel bodily to leeward.

Leeway angle* the angular difference between the water track and the ship's heading.

Ded reckoning the process of maintaining or predicting an approximate record of progress by projecting course and distance from a known position. Derived from deduced reckoning . Also called 'dead', loosely.

DR position (DR) (1) the general term for a position obtained by ded reckoning; (2) specifically, a position obtained by using true course and distance run.

Estimated position (EP) a best possible approximation of a present or future position. It is based on course and distance since the last known position with an estimation made for leeway, set and drift; or by extrapolation from earlier fixes.

*The word *angle* will be omitted in normal use unless there is a possibility of confusion.

This chapter discusses the skills needed to determine a yacht's position after several hours' sailing without sight of land. Such a position is known as an *Estimated Position* (or EP) and is shown on the chart as a small triangle. As defined above, it represents the navigator's best attempt to show the vessel's position, having allowed for tidal stream, leeway and any current. Although it should never be wildly inaccurate, it will generally have a low order of absolute accuracy — especially after several hours' sailing in strong winds and tides. The term 'sailing' is used advisedly because this procedure is of particular importance to the navigator of a sailing yacht, when beating to windward; the reason being, that when beating, the course has to be changed with each shift in wind direction as well as when tacking. However, *all* craft are affected by tidal streams, leeway and current. The final exercises in this chapter will deal with this situation, which can also occur when under power. It is, however, a basic skill that all aspiring navigators must acquire.

The EP is arrived at via the vessel's *Dead Reckoning* position or DR. A dead reckoning position is wholly theoretical and represents where the vessel would be *in the absence of wind and tide*. It is always plotted as a result of two fundamental measurements: *direction* steered via the compass and *distance sailed* from the log. It therefore follows that the accurate calibration and full understanding of both these vital instruments is very important, as has already been mentioned in Chapter 5.

Note: In the interests of clarity and simplicity, please note that the word COURSE is used in the place of Water Track (or Wake Course as it was known until recently) in this text.

We will, therefore, have three distinct and different words for the three possibilities.

> Heading — the ship's longitudinal axis
> Course — the ship's path through the water.
> Track — the ship's path over the ground.

Government approved terminology will be shown in brackets when required.

Plotting terms and navigator's symbols

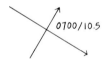

A two-bearing fix (probably from a hand bearing compass) taken at 0700 when the log read 10.5 M.

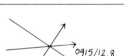

Excellent three-bearing fix, taken at 0915, when log read 12.8 M.

Less accurate but acceptable three-bearing fix, producing a small 'cocked hat'. Time 2040, log 76.4 M.

Latitude and longitude position, perhaps obtained from Decca or transferred from another chart.

1421/10.9	Fix from a range and bearing, possibly from a radar measurement.
	Transferred position line.
060°T 7	Course through the water 060° True. Speed through the water 7 knots.
0700/14.2	DR position, calculated at 0700 when log read 14.2 M.
8	Track (ground track) or path taken relative to the earth's surface. Ground speed 8 knots.
	Tidal stream vector (set and drift).
△ 0810/14.4	Estimated position for 0810. Log 14.4 M.

The Triangle of Velocities

This concept is at the very heart of navigation and a firm grasp of the principles involved is essential. First, the navigator must understand that, in a tidal stream (or in a current), the vessel for which he is responsible will have two simultaneous velocities, one relative to the sea and one relative to the earth. The cause is a third vector quantity (set and drift) due to the tidal stream (the *horizontal* water movement caused by tides). His need is to know his path and speed relative to the earth, which we call *track and ground speed*. However, the basic instruments (compass and log) supply information *relative to the water*. These two quantities are simply known as *course* and *speed*. To turn course and speed into ground track and ground speed, we plot two sides of a triangle. We begin with a line on the chart, starting from a known position, which represents the course steered (from the compass) and the distance covered (derived from two log readings). Next plot the tidal stream that the vessel is assumed to be experiencing. This vector, representing set and drift, must be proportional in length to the time involved. For example, should the time between the two log readings be 40 min or two-thirds of an hour, then the drift must be plotted as two-thirds of the tidal rate.

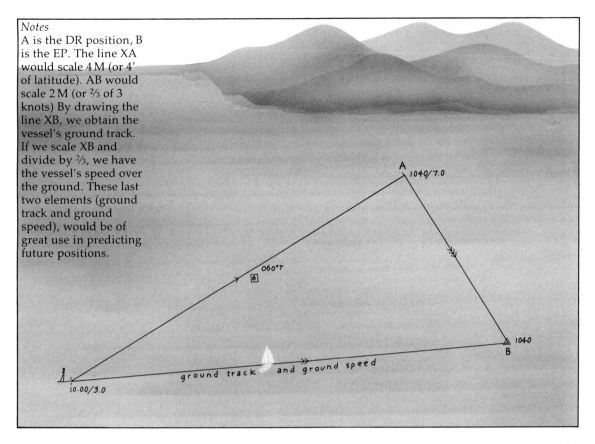

Notes
A is the DR position, B is the EP. The line XA would scale 4 M (or 4′ of latitude). AB would scale 2 M (or ⅔ of 3 knots) By drawing the line XB, we obtain the vessel's ground track. If we scale XB and divide by ⅔, we have the vessel's speed over the ground. These last two elements (ground track and ground speed), would be of great use in predicting future positions.

Example *Data:* ship's position was X at 1000 when the log read 3.0 M. Course 060° True. Tidal stream predicted at 135°, 3 knots. Plot the estimated position (EP) for 1040, when log reading was 7.0 M.

The following exercises are carefully graduated in difficulty and are used to introduce the important concept of *leeway*.

Exercise 1

Chart 5043
Plotting DR and EP over *one* hour.
At 0800, log 4.0 M, your position is 51°25′.00N, 1°30′.00E. Course 040° True (use outer compass rose).

a State the latitude and longitude of the vessel's EP for 0900 when the log reads 10.0 M. Tidal stream is 120°, rate 1.8 knots.

b Give the track and ground speed.

c Apply full conventional coding to your plot.

Exercise 2

(continuation from Exercise 1) Plotting DR and EP over 1¼ hr. From your 0900 position, lay off a course of 160° T and work up an EP for 1015. Log reads 17.5 M. Assume a tidal stream of 2 knots at 114°.

a What is the lat. and long. for the 1015 EP?

b What is your ground speed and ground track?

Exercise 3

Stanford chart
In Exercise 3 we are going to work up an estimated position over 2½ hr. This will, of necessity, involve three tidal stream vectors, one for each hour or part of an hour. It is important to realise that the three vectors can

be laid off cumulatively, thereby saving a great deal of time. The diagram should make the procedure clear.

Time 1300. Position 1 M to the E of the Platte Fougère light house. This light will be found 1 M NE of Guernsey. Course 070° T, log 4.0 M.

a Give the lat. and long. of your position at 1530, log 19.0, with the following tidal stream data:

1st hr 1300−1400, 030° at 1.8 knots	
2nd hr 1400−1500, 005° at 2.5 knots	
3rd hr 1500−1600, 340° at 3.0 knots	

b What was your track and ground speed in the last half hour? *Note:* to do this accurately, first plot the DR position for 1500. The log reading for 1500 will be 16 M, as the vessel is proceeding at 6 knots (15 M in 2½ hr). Plot the DR 12 M along the course line ⟶ . Next, transfer with a parallel rule the first and second tidal stream vectors (hours 1 and 2): this will give the 1500 EP. To obtain track, join EP for 1500 to EP for 1530. *EP to EP, or fix to fix, always produces a track.*

c Is the vessel on course to clear Cap de la Hague? *Note:* to do this, produce the track line.

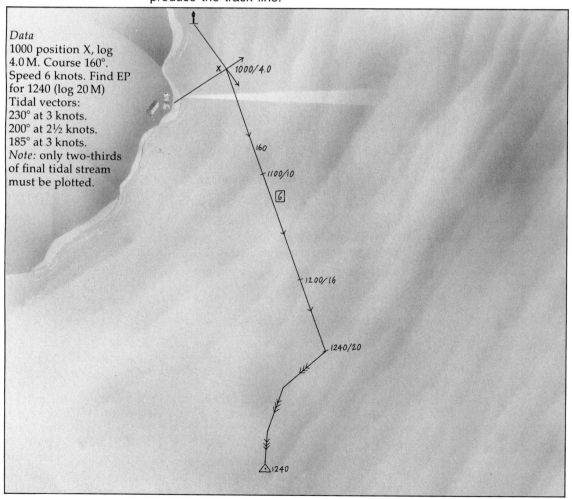

Data
1000 position X, log 4.0 M. Course 160°. Speed 6 knots. Find EP for 1240 (log 20 M) Tidal vectors:
230° at 3 knots.
200° at 2½ knots.
185° at 3 knots.
Note: only two-thirds of final tidal stream must be plotted.

Working up DR and EP after course changes
The point of the exercise is to show that course changes do not affect the issue; the same cumulative procedure can be adopted. If a specific track is required, then the two relevant EPs must be plotted, as in the previous exercise.

2⅔ hr plot with two course changes
Data time 1400. Position Sardrière buoy (2½ M SE of St Peter Port). Course 205°T. Log 2.2 M. At 1452 (log 7.3 M) the wind veers and allows the vessel to steer 220°T. At 1535 (log 10.5 M) another veer occurs and 230°T is steered.

a Plot the EP for 1640, log 16.8 M. State the position as a bearing and distance from Les Hanois light house (SW Guernsey).
Tidal stream data

1st hr 1400−1500, 290° at 1.8 knots	
2nd hr 1500−1600, 260° at 1.2 knots	
3rd hr 1600−1700, 240° at 1.2 knots	

A good method of checking the overall length of the course (fourth log reading minus first log reading i.e. 14.6 M), is shown in the second diagram in Chapter 1. This is well worthwhile, as it is both a quick and easy way of checking several subtractions.

b What was the average track and ground speed over the 2⅔ hr period?

Leeway

Before going any further we must now develop the idea of leeway and the closely allied terminology connected with sailing a yacht to windward. As anyone will know who has manoeuvred a boat (sail or power) within the confines of a marina on a windy day, boats do not necessarily go in the direction in which they are being pointed. This tendency to drift sideways, downwind, is called leeway and must be borne constantly in mind when navigating. Sailing vessels beating to windward in rough conditions will make at least 10° leeway, and more if bilge-keeled or lacking in draft. Powered craft also make leeway, depending on the ratio of their superstructure to draft. Leeway is accentuated by low speeds, breaking seas and high winds. Indeed, if high wind persists, a wind-induced current is created on the sea's surface and this must also be allowed for as leeway.

Assessing leeway

This is a difficult subject and one that responds to experience, especially with one's own boat. Some good may be done with careful back bearings (of an object astern) but it is often hard to distinguish between drift, which is due to tide, and leeway due to wind. One method is to drop a page of a newspaper overboard and note the bearing before it disappears. The wet paper has the advantage of not being blown to leeward, but the disadvantage of not distinguishing wind-induced surface current. A comparison of the reciprocal of the ship's heading with the bearing of the paper will reveal the appropriate leeway. In the absence of tidal stream, a Decca Navigator will enable leeway to be measured accurately. Simply compare the compass heading (True) with the 'course' being displayed on the Decca set. Make sure you only take readings after the averaging time (pre-

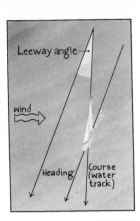

entered into the set) has been sailed in a straight line. (*Note:* this measurement can only be carried out successfully at slack water.)

Allowance for leeway

Difference between Heading and Course
Set and drift due to tide are allowed for as a vector quantity, but leeway is allowed for as a difference between the angle formed by the ship's heading (fore-and-aft axis) on the one hand and her course (water track) on the other. These two directions are best described by two distinct words, namely *heading* and *course*.

Basic rule
The course must be downwind (to leeward) of the heading, *or* heading must always be to windward of course (water track).

Exercise 5

a What heading must be steered in order to achieve a track of 060°T? Wind is northerly and leeway assessed at 12°.

b The wind is SW and the yacht is being sailed hard on the wind on a heading of 175°T with leeway 10°. What course must be laid off on the chart?

Some sailing terms related to navigation

Tack An abrupt change of heading as the sailing vessel zig-zags into the wind, usually about 90° on the average yacht.

Port and starboard tack A vessel is said to be on starboard tack when her starboard side is presented towards the wind.

Closehauled Vessel is being sailed at a minimum angle to the wind, usually about 40°−45°. This produces a situation of maximum leeway and also one in which the course is being dictated by the wind direction. Accurate navigation of a yacht sailing closehauled offshore is very demanding.

Reaching A term used when the wind is approximately at right angles to heading. Some leeway, but good speed.

Running The yacht is 'before the wind', i.e. with her stern to it, and often under spinnaker. Negligible leeway.

Leebowing the tide This is the situation skippers must try to produce when beating to windward, although it does make for rough sea conditions. The vessel is said to be lee-bowing the tide when the tidal stream is 'lifting' her into the wind, i.e. the tide is on the lee bow and is increasing the apparent wind and tending to set her in the desired direction of travel.

Luff up To turn towards the wind.

Bear away Opposite to luff up.

VMG or V_mg	Velocity made good to windward. Speed achieved in a straight line measured directly into the wind. E.g. speed logged 7 knots, tacking angle 90° allowing for leeway, VMG 5 knots.
Wind backing	Shifting anticlockwise.
Wind freeing	Its direction moving aft, i.e. it comes more from the stern.
Veering	Opposite to backing.
Headed	Wind shifts so it comes more from ahead, forcing the vessel to bear away.

Exercise 6

Part I

A yacht is being steered on a heading of 300°T. The wind is westerly and leeway 8°.

a Which tack is she on?

b Is the vessel running, beating or reaching?

c What is her course?

Exercise 6

Part II

A yacht is closehauled on the starboard tack. Her intended course is 126°T, leeway is 12°.

a What heading must be steered?

b What is the approximate wind direction?

c Would the skipper prefer the wind to back or veer?

Important note: When plotting a situation in which leeway is involved, *never plot the heading.* Leeway and headings may be noted in the ship's logbook but *never* put on the chart. If you refer to the navigator's symbols (above) there is no symbol for heading; only courses and tracks are illustrated.

Exercise 7

Stanford chart

A short beat to windward, allowing for leeway. Position 1 M E of Platte Fougère (NE Guernsey) Time 1300, log 4.1 M.

   The yacht is closehauled on the starboard tack, steering a heading of 330°T. Leeway 10°. At 1430, log 12.6, the yacht tacks and is able to steer a heading of 060°T.

a Give the latitude and longitude of her estimated position at 1530, log 17.8 M. Assume 10° leeway throughout and the following tidal streams:

1st hr 1300–1400, 065°T at 1½ knots

2nd hr 1400–1500, 045°T at 2.0 knots

3rd hr 1500–1600, 030°T at 2.8 knots

b From the steering compass, through what angle did the yacht tack? (No leeway allowance.)

c Through what angle did she tack through the water? (Include leeway.)

d Decide if she had a lee-bowing tide on neither tack, one tack or both?

e What was the approximate wind direction?

Stanford chart

A longer beat to windward. The following is an extract from a navigator's log, as a yacht attempts to beat NE towards Cap de la Hague. Initial position at 0643 is Noir Pute, between Herm and Sark.

Time	Log	Heading	Leeway	Course	Remarks
0643	19.7	073°	10°	?	Wind NNE Force 5
0715	22.5	086°	12°	?	Rough, wind veering
0733	24.0	005°	10°	?	Tacked
0830	30.2	017°	8°	?	Wind veering NE
0918	34.8	110°	10°	?	Tacked
1000	38.5	110°	10°		

Tidal streams:

1st hr 0630–0730, slack

2nd hr 0730–0830, 270° at 1 knot

3rd hr 0830–0930, 225° at 2 knots

4th hr 0930–1030, 180° at 2.2 knots

a What is the lat. and long. of the 1000 EP?

b What is the approximate VMG between 0643 and 1000?

c Looking at the tidal stream data, is the VMG likely to improve, remain constant or worsen in the next few hours?

Stanford chart

Finding the tidal stream from compass, log and two known positions. *Note:* the difference between a dead reckoning position (DR) and a fix (or EP) is due to the tidal stream. This exercise is placed N of Jersey. Course 270°T, time 0200, log 14.70. Position: Sorel Point Light bears 156°T (Jersey N coast). Grosnez Light bears 192°T (NW Jersey). Desormes buoy bears 240°T (4 M NNW of Grosnez). At 0315, log 22 M, the position by Decca is 49°22'.6N, 02°26'.7W.

a What is the rate (in knots) and set of the tidal stream?

b Fully code your plot.

c What was the track and ground speed?

Summary

Winds are always stated in the direction *from which* they appear to originate.

Heading	Longitudinal axis of vessel, i.e. the compass reading.
Course (Water track)	Path taken through water, i.e. heading plus or minus leeway.
Track (Ground track)	Path taken over the ground.
Speed	Speed through the water, i.e. speed being logged.
Ground speed	Speed over the ground.
DR	Position as determined by projecting course and distance from a known position.
EP	Position calculated from all available data.

Note: Electronic systems indicate tracks and ground speed.

Factors determining leeway

Wind — Increasing wind, increasing leeway.

Wind angle — Maximum leeway for sailing vessels when closehauled. Maximum leeway for power vessels when apparent wind is abeam.

Wind duration — A strong wind of some days duration creates a surface current.

Breaking seas — Breaking seas lift small vessels to leeward.

Draft — The less the draft, the greater the leeway.

Freeboard and superstructure — The greater the windage, the greater the leeway.

Speed — The slower you sail, the greater the leeway you make (this explains why sailing boats suffer most when closehauled). Heading is always to windward of course. Course is always to leeward of heading. *Never plot headings.*
Note: always associate ground speed with track and speed with course (water track).

Answers to Chapter 6

Exercise 1
a EP at 0900 51°28'.70N, 01°38'.60E
b 056°T, 6.6 knots.
c Time and log reading at starting point.
Single arrow on course line.
040° written near arrow.
Time and log reading at DR position.
Triple arrow on tidal vector.
Triangle and time at EP
Double arrow on track.
Speeds in boxes near appropriate vector.
Note: on all answers derived from chart plotting, any answer, particularly one derived at the end of a lengthy plot, will have a slight error. All such answers must, therefore, be regarded as approximate and for that reason are only given to a realistic place value, i.e. 1° on courses and 0.1 M on distances.

Exercise 2
a Approximately 51°20'.70N, 1°46'.30E
b Ground speed 7½ knots, track 148°.
Notes on some possible errors:
2a: check that tidal stream vector is 2½ M not 2 M.
NB 2 knots for 1¼ hr gives a drift of 2½ M.
2b: ground speed is not 9.36 knots. Again allowance must be made for the hour and a quarter.
9.36 M ÷ 1.25 hr = 7.5 knots.

Exercise 3
a 1530 EP 49°41'.50N, 02°04'.80W
b Track 043°T, ground speed 6.8 knots
c Yes. Clearance 2 M off lighthouse.

Exercise 4
a 198° from Les Hanois light, 11.8 M. If slightly in error, check that you have taken ⅔ of final tidal stream vector.
b Mean track 228°, ground speed 6.2 knots.

Exercise 5
a 048°T. *Note:* winds are always described in terms of the direction *from* which they originate.
b 165°T.

Exercise 6 Part I

a Port.

b Beating.

c 308°

Exercise 6 Part II

a 138°T.

b Southerly.

c Veer. Veering would produce a 'freer' wind, or a more southerly course could be pursued.

Exercise 7

a 1530 EP. 49°42′.2N 02°23′.0W

Should you be in error, check the following: initial course should be 320°T, second course should be 070°T, third tidal vector should scale 1.4 M.

b 90°

c 110°

d Lee-bowing tides on both tacks

e 015° or approximately NNE

Exercise 8

a EP for 1000 49°34′.7N, 02°17′.1W

Checklist in case of error: total distance logged 18.8 M, total tidal stream drift 4.0 M (not 5.0 M); courses 083°, 098°, 355°, 009° and 120°.

b Distance gained to windward. Noir Pute to 1000 EP is 8.1 M. Time taken 3 hr 17 min or 3.28 hours. VMG = 8.1/3.28 = 2.47 knots or about 2½ knots.

c Worsen. The ebb tide has not yet run its course. A start should have been made roughly 6 hours earlier or 6 hours later.

Exercise 9

a Tidal stream 315° at 1.76 knots (2.2 M ÷ 1.25 hr)

b Check on coding of chart plot.

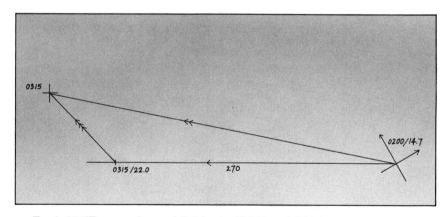

c Track 281°T, ground speed 7.4 knots (9.3 M ÷ 1.25 hr)

Position Fixing

Position fixing is an activity which keeps the conscientious navigator busy, especially when cruising a foreign coastline or when beating to windward. This preoccupation is not only with the yacht's position but also her future track, which may be determined from two fixes a few miles apart. The main sources of information in these circumstances will be position lines from bearings taken with a hand bearing compass. To do this successfully, the following criteria must be met:

1 The hand bearing compass must be used in a deviation-free position aboard the boat.

2 At least two, or preferably three, objects to be chosen which can be identified visually and on the chart.

3 The angles between the bearings of any two objects must be greater than 30° but not more than 150°.

4 Given the choice, a nearer object should be chosen in preference to a distant one.

All these criteria are concerned with accuracy. In any measurement there will always be a degree of error. In the case of a hand bearing compass this will be between ±1° and ±5°, depending on how rough the sea is at the time. Small angles of 'cut' between position lines and those derived from distant objects increase the area of uncertainty (see diagrams). For an object 20 miles away, the area is increased by ⅓ n. mile for each degree of error. This would halve if the object were only 10 miles distant.

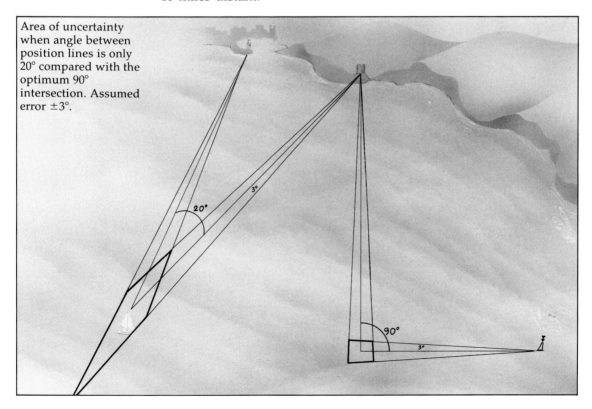

Area of uncertainty when angle between position lines is only 20° compared with the optimum 90° intersection. Assumed error ±3°.

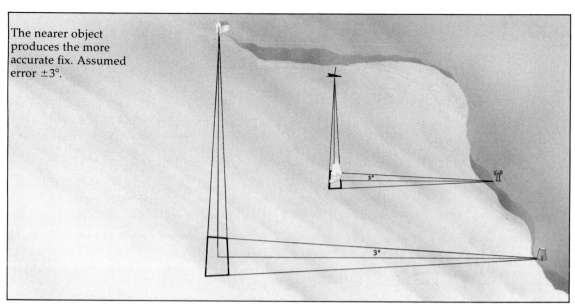

The nearer object produces the more accurate fix. Assumed error ±3°.

Using chart 5043 and an assumed position of 51°20'.0N, 1°30'.0E.

a Explain why Broadstairs Knoll buoy would be preferable to North Foreland Light for fixing one's position.

b Explain why Gull buoy and Goodwin Knoll buoy could not both be used as the source of two more position lines.

c Explain why Grandville Hotel and the North Goodwin Light Vessel could not be used in conjunction with each other.

Using a hand bearing compass

1 Under rough conditions, patience is needed in taking bearings. If possible, two observers should be used and a mean value plotted.

2 Use transits whenever possible. They are error-free and can be pre-plotted. One other well chosen bearing will then produce an accurate fix.

3 Use the echo-sounder as a *possible* confirmation of position when only one or two position lines are available.

4 As bearings change with time, especially if the yacht is sailing fast, take the beam bearing last, as this will be the one subject to the most rapid rate of change of bearing.

5 When laying off a course from a three-bearing fix that has produced a 'cocked hat', choose the point closest to danger.

Exercise 2

Chart 5043

a At 1004, North Goodwin Light Vessel bears 310°M and the East Goodwin buoy (Q (3) 10s) bears 220°M (variation 5°W). By 1041 the bearing of the East Goodwin buoy has changed to 290°M and the East Goodwin Light Vessel bears 228°M. What was the yacht's track and ground speed between these fixes?

Exercise 3

Chart 5043, clearing bearings
You intend sailing SW between the Goodwin Sands, using Kellett Gut.

a What would be the safe minimum and maximum bearings on the East Goodwin buoy?

74

b Which other instrument would be of great assistance while navigating Kellett Gut?

The Sextant

This instrument, capable of measuring angles to less than a minute of arc, can be used in coastal navigation both horizontally and vertically. As this practice is now very little used and requires either a special chart-plotting instrument or special tables, no exercises will be given in horizontal angle fixing or distance off by angle of elevation.

Fixes from a mixture of position lines

It is surprising how inhibited many navigators are of combining two or three methods in order to achieve a fix. Mixed methods are especially useful when only one identifiable object is in sight. Having laid off the position line on the chart, a reasonable solution to determine the approximate distance off would be to plot the EP for that time and take a position along the position line closest to the EP. Alternatively, a radio bearing could be taken and crossed with the visual bearing. In both cases confirmation may be forthcoming from the echo-sounder. Occasionally a series of soundings can be taken and matched to the chart to give a position. If this technique is employed, then the following points have to be observed:

1 The soundings must be 'reduced'; that is a deduction has to be made which compensates for the present height of tide above datum. (see chapter 3)

2 The soundings, at appropriate intervals, have to be written along the edge of a piece of paper.

3 The paper has to be laid on the chart in the direction of the vessel's probable track and matched, if possible, to the soundings and contour lines on the chart.

<table>
<tr><td>Exercise 4</td></tr>
</table>

Line of soundings

a Using the Stanford chart find the probable lat. and long. of the yacht at 1410. She is on passage from St Peter Port in Guernsey SE towards St Malo. Off Jersey she encounters fog. At 1340, approaching the Minquiers, the following soundings were taken at 5 minute intervals:

Time 1340: 25 fathoms   Time 1400: 15 fathoms
     1345: 24   1405: 12
     1350: 24   1410: 9
     1355: 24   1415: 16

These soundings need to be 'reduced' by 4 fathoms (see data below).

Ground speed 6 knots (or 1 M every 10 min). Track 140°.
Height of tide at 1400 approximately 7.5 m (approx. 4 fathoms).

b What course alteration must be ordered at 1415?

We must now tackle the problem of fixing position when only one object is in sight. Typically this happens most often at night when only one light is visible. The difficulty is one of range. Apart from a radar set or a sextant, there are no direct ranging devices available which are satisfactory over 3 miles. If, however, the light is seen just touching the horizon, the range can be determined via Dipping Distances Tables which are published in all nautical almanacs. To carry out this procedure the following information has to be available:

1 Observer's height of eye above sea level.

2 Height of light (available from chart or Light List but stated relative to mean high water springs MHWS). (*Note:* height of lights is given in metres (small m); capital M is range in nautical miles.)

3 Sea level relative to MHWS.

4 Table of dipping distances

Example 1 Finding the distance off.
At 0200 the Casquets Light is seen just rising and dipping on a bearing of 130°T. Height of eye 2 m, height of light 36 m (approx. 120 ft). Height of tide at Alderney at 0200, 2 m. MHWS Alderney 6 m. Therefore, height of light is 40 m (6 − 2 + 36). Distance off (from tables) is 16.1 M.

Exercise 5 Dipping distance Chart 5043
a What is your distance off the North Goodwin Light Vessel, if the light is seen rising and dipping? Height of eye 2 m.

b Give the latitude and longitude of your position at 0340, when the North Foreland Light is seen rising and dipping on a bearing of 328°M. Height of eye 3 m. Height of tide 2 m. *Note:* MHWS North Foreland can be found at the southern edge of the chart, using Ramsgate or Margate data.

An interesting method of fixing position from a single object is to take two bearings on it at different times; they must differ by at least 30°. The method is known as a 'running fix' and is sometimes very useful. It has only a modest degree of accuracy, as a large amount of data has to be plotted each with its own inherent error. Several exercises involving running fixes are given because it is also an excellent means of revising many of the skills already covered.

Example 2 Running fix
Time 1100. Log 6.0 M. Course 060°T. Light bears 090°.
At 1130, log 9.0 M, light bears 160°. Tidal stream 010° at 3 knots.

Note: the point at which the course is laid off from the first bearing at 1100 is not critical. In practice, one would estimate the distance off by eye or use an EP.
The initial plot would be as follows:

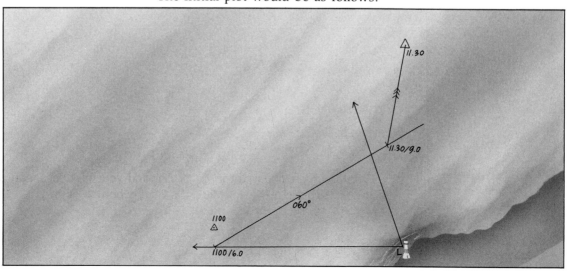

In this diagram the plot has been completed by *transferring* the initial bearing AL, with a parallel rule, to the estimated position E. Where the transferred position line TE cuts the second bearing BL, we obtain our fix at F. Remember:

1 A fix is obtained where two position lines cross (TE and BL).

2 The symbol for a transferred position line is ——▸—▸

3 The tidal vector DE must be plotted as 1.5 M, not 3 M.

4 The start of the plot is at A, but the position of A is not critical.

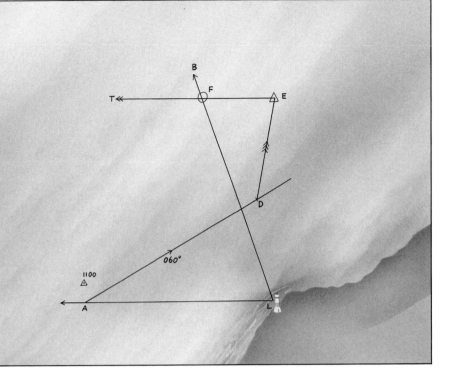

In this example, the 1130/10 DR falls beyond the second bearing; in later examples it may fall short of it or even on it. The position of the 1130 DR relative to the second bearing has no significance.

The next diagram shows the same plot completed. To help the reader, all points have been labelled for future reference.

Exercise 6

6a Plot Example 2 on chart 5043, using the South Falls buoy (51°14'.0N, 1°44'.0E), as the single visible object. State your running fix position at 1130 as a bearing and distance from the South Falls buoy. I suggest that the course is laid off from the initial bearing at approximately 2.5 M from the buoy.

6b Repeat exercise 6a, commencing the plot 1.5 M from the buoy.

Now look again at the diagram above. If we put in line AE we have the yacht's track (044°T) but it is in the wrong position. If we transfer it to our fix (at point F) we have a more correct picture. If we project it back to our initial bearing, we find that our position at 1100 was, in fact, just under 4 miles from the South Falls buoy. If we produce (extend) it forward, we see that we are on track to pass just W of the Mid Falls buoy. Thus we see that by picking up our track from the triangle of velocities (ADE) and transferring it to our fix at 1130, we can forecast future positions. Such forecasts must not be carried too far forward in time without thinking about changes in data, especially the tidal stream rate and direction.

6c From your previous plot in 6a or 6b, find your ground speed and give the approximate ETA for Mid Falls buoy when abeam.

It is now a matter of further practice on the Stanford chart. In these exercises, a more realistic approach will be used.

1 Objects really will be the only ones in sight, given average visibility.

2 In some cases an unlit hazard will have to be cleared, thus creating a real need for such a fix.

3 Leeway will be re-introduced. Remember: 'Heading is always to windward of course (water track).'

4 Compass errors will be included. Remember: CADET, CDMVT and that deviation is heading dependent.

Exercise 7

Stanford chart

At 0310 the Grand Léjon Light (48°45′.0N, 02°40′.0W) is the only light visible on a bearing of 228°M, at an estimated range of 4 miles. Course is 294°C (use *Zephyr's* deviation card, chapter 5). Log at 0310 reads 22.6 M. At 0350, log 27.7 the light bears 160°M. Tidal stream is calculated at 070°T at 1.8 knots.

a What is your distance from the Grand Léjon Light at 0350?

b What is your track and ground speed?

c How far off the light were you at 0310?

d Given a constant tidal stream, water speed and course, are you on track for the Horaine Light?

e What is your ETA at the Horaine Light?

Exercise 8

Stanford chart

This running fix is off the Roches Douvres, 49°06′.5N, 02°49′.0W. As the chart shows, the light structure, an impressive 197 ft high, is on the eastern side of the reef. When wishing to pass to the west, especially at night, one must ensure that the yacht's track remains not less than 3 miles clear of the light. In the absence of radar or Decca, it is the classic justification for a running fix, backed up by regular reading of the echo-sounder.

a What is your distance off the Roches Douvres Light at 0407 given the following data? Time 0250, log 73.4 M, light bears 209°M. Heading 250°C, wind NW, leeway 8°. Estimated distance off 7 M. At 0407, log 81.6, light bears 175°M. Tidal stream 100° at 2.0 knots.

b By how much will your projected track clear the westernmost reefs?

c What is your track and ground speed?

Note: in this somewhat tight situation and in view of the low order of accuracy inherent in a running fix, it would be prudent to luff up a further 10° to a heading of 260°C if the wind allows. Failing this, you would have to consider tacking. A brave man might consider standing on but would

a Rouse all hands and be prepared to tack.

b Double the lookout.

c Have a constant watch kept on the echo-sounder.

Exercise 9

List the measurements required for a running fix and appraise their accuracy on an A to C scale:

A High degree of accuracy

B Good accuracy

C Acceptable accuracy.

Assume all instruments have been correctly calibrated and that sea conditions are smooth or slight.

Doubling the angle on the bow

This is a simple variation on the running fix theme. The scenario is rough conditions and a yacht that is being sailed either shorthanded or singlehanded. It depends on the well known properties of an iso-sceles triangle, and has the great advantage of not requiring a visit to the chart table. The basic idea is to note the log reading when the light has a bearing relative to the bow of some X° (X being not less than 30°, and also noted). Sailing on in as straight a line as possible, the lone yachtsman notes the log reading when the relative bearing on the bow becomes 2X°. The distance off the light must be equal to the distance sailed. There is, however, one big snag: the distance sailed must be measured *over the ground*, and unfortunately the log readings will give distance sailed *through the water*. In the absence of tidal stream, therefore, the system is both sound and practical. With a foul or head tide, the distance off will be underestimated, which is potentially dangerous. Under these circumstances the helmsman would have to take a quick look at the tidal atlas and make a mental adjustment to his distance run. With a favourable tide the tendency will be to under-estimate the distance off, which is usually not too serious unless other hazards are around.

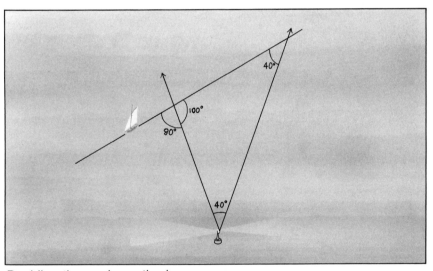

Exercise 10

Doubling the angle on the bow
At 0200 you are approaching Roches Douvres reef from the N, intending to pass well to the W. You are sailing singlehanded before a strong wind from the NW, and therefore unable to leave the tiller for more than a few seconds at a time. Your course is 225°M and the light bears 185°M; log reads 20.4. By 0240 the bearing on Roches Douvres is 145°M and the log reads 24.5. A quick glance at the tidal atlas shows the stream to be slack and then running W after 0300. You know that the reef extends at least 2 miles to the W of the light.

a Without plotting on the chart, calculate how far off the light you are at 0240 (a rough sketch is permitted to help you envisage the situation).

b On which tack is the yacht?

c Do you need to bear away, luff up or stand on?

Summary

Type of fix	Accuracy
Three-bearing	Very good (accuracy can be assessed)
Two-bearing	Good
Running fix	Acceptable
EP	Acceptable, but depends on the length of time since last visual fix and on weather conditions.
Doubling angle on the bow	Acceptable in the correct context.
Radar	Good. Needs experienced operator. Positive object identification difficult. Radar range more accurate than bearings.
Decca	Very good. Updated every 20 sec. Accuracy decreases with range (200 M) Less good at night and in winter. Has potential accuracy indication.
Radio direction finding	Good as homing device. Best fixes only made with patience and practice.
Lines of soundings	Acceptable, but seldom practical.
Dipping distance	Good in calm conditions.
Horizontal angle fix with sextant	Very good
Distance off by vertical angle (sextant)	Good, especially over short ranges.
Satellite receiver (Satnav)	Very good, with coverage world-wide, but 1–2 hr intervals between fixes.

Answers to Chapter 7

Exercise 1
a   The buoy is the nearer.
b   Angle subtended is less than 30°.
c   Angle subtended is greater than 150°.

Exercise 2
a   Track 166°T. Ground speed = 3.2 M in 37 min or 5.2 knots.

Exercise 3
a   055°M to 061°M approx. *Note:* 055°M does not clear the 2.7 m wreck, 8 cables SW of the buoy. Buoy may disappear from sight at a range of 3 to 4 miles.
b   Echo-sounder

Exercise 4
a   48°56'.5N, 2°19'.2W
b   90° to starboard

Exercise 5
a   10.1 M
b   51°06'.6N, 1°45'.4E
Height of light = 57 m above MHWS
MLWS about 5 m
Height of tide = 2 m
Height of light at 0340 = 60 m
Distance off = 19.7 M

Exercise 6

a 340°, 3.1 M

b 340°, 3.1 M

c 1206 hours. Distance 5 M. Ground speed 8.25 knots. Duration 36.4 min.

Exercise 7

a 4.3 M. Course 291°T. Drift 1.2 M (⅔ of 1.8 M).

b Track 304°T. Ground speed 6 knots (4 M in 40 min).

c Distance off 2.2 M.

d Yes. But a little too close: should alter course 5° to starboard.

e 9.5 M to go, at a ground speed of 6 knots will take 95 minutes or 1 hr 35 min. Therefore ETA = 0525.

Exercise 8

a 3.0 M

Course true = 235°

Logged distance = 8.2 M

Drift 2.6 M $\left(\dfrac{2 \times 77}{60} \right)$

b 1.1 M

c 218°T (compare this with your heading of 250°C!) Ground speed 5.2 knots $\left(\dfrac{6.7 \times 60}{77} \right)$

Incidently, if you made the track 222°, the error is to have used the equivalent of AF in the diagram instead of AE.

Exercise 9

2 bearings, A grade of accuracy

2 log readings, B grade

Compass course, C (B if steered by good autopilot)

Tidal stream data (set and rate), C.

Note: unfortunately, tidal stream data has only a modest reputation for accuracy. Too often it has been extrapolated from too few observations of too short a duration. The wind also affects surface current.

Exercise 10

a 4.1 M

b Starboard

c Stand on. You should have a clearance of 2 M (or 100%) and the tide will increase this. Watch cockpit echo-sounder, which must not read less than 30 fathoms (54 m).

Chapter

8

Setting Course

A yacht on a correctly calculated course is a safe yacht. A correct course infers not only a safe present position, but that future positions will be equally secure and a good landfall will eventually be achieved. Sailing the correct course requires that: (1) the compass should be free of error or its errors well known and allowed for, and (2) tidal streams are correctly assessed. Both these topics have been dealt with in some detail in previous chapters. It now remains to combine these two factors successfully into their final product, namely, the correct course to steer. On many occasions course setting is not a problem because either the tidal streams are almost nonexistent or they run parallel to the yacht's path, often the case when following a coastline. Under such circumstances course and track are the same: the tidal streams only affect the estimated time of arrival. It is when crossing channels or sailing between islands or hazards, especially in poor visibility, that good course setting becomes a skill well worth acquiring.

Speed

The slower a boat, the more she is at the mercy of the tidal streams. For this reason, navigation is simplified by motoring or motorsailing across strong tides in poor visibility. This will not only increase your speed but it will also give two other valuable attributes, consistency and predictability. As we shall see, successful course setting requires the navigator to forecast the yacht's speed through the water correctly. When making one's way in a series of short hops, from headland to headland or buoy to buoy perhaps in fog, a constant ground speed of 6 knots is of enormous help. At 6 knots the boat is covering the ground at the rate of a cable every minute, and this means that a series of ETAs can be obtained without calculation.

Exercise 1

Chart BA 5043
You are on passage northwards through the Downs (51°13′N, 1°26′E) at 6 knots ground speed. At 0840 BST you have Goodwin Fork buoy close abeam. What is your ETA at the following buoys? Downs, South Brake, NW Goodwin and Gull Stream.

Two modes of course setting

We will eventually tackle setting course in two stages. Short term course setting deals with a time span of approximately 1 hr or less and usually needs only a single tidal stream vector. Long term course setting requires multiple vectors, one for each hour or part of an hour.

Revision
First it may be necessary to revise the triangle of velocities, introduced in Chapter 6, as being at the heart of navigation. The triangle consists of three vector quantities and therefore embodies six measurements, three directions and three speeds. It was initially used to establish the yacht's EP, via her DR. A typical plot may look like those overleaf.

The initial fix A is obtained at 1120 (log 10 M). The DR position an hour later (1220, log 16 M) is at B. The SW stream is then plotted to find the yacht's EP at C. The path over the ground during the hour is the dotted line AC, the vessel's ground track. An obstruction along the line AB would not present a hazard; indeed, on some plots AB may be partially across dry land. The speed over the ground may also be discovered from scaling AC. It is important to realise that ground speeds must always be applied along track lines, and that speeds (by which we mean speed through the water) must invariably be applied to the course line. An extension of the line AC will be a prediction of future positions after 1220 if the tidal stream remains the same.

In this plot, we have a situation in which the course is known and the need is to find where the yacht is going, i.e. her track. This situation is unusual and will only occur regularly when a sailing yacht tacks to windward. The more normal situation is to know the intended track and to need to find an appropriate course to steer in order to attain this track. In other words, in most sailing situations the navigator has to calculate the course in order to sail from departure point to destination. To do this he usually draws a slightly different triangle of velocities, in which the tidal stream is plotted on the departure point. First, let us see what information he has available and which information he needs to calculate.

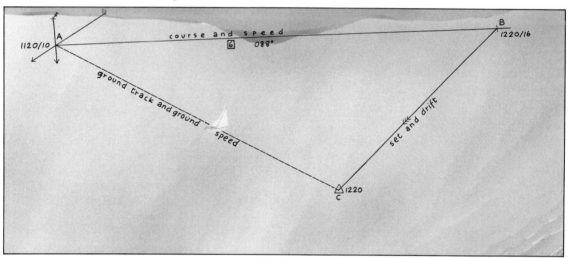

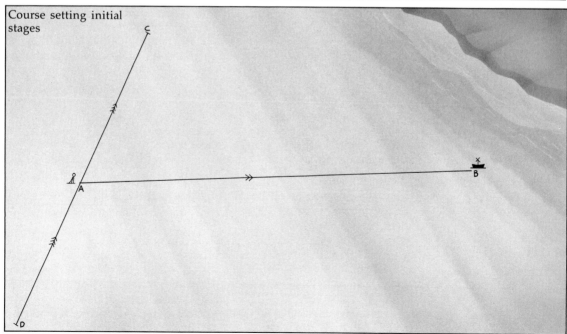

Course setting initial stages

In a course setting, one needs to know at least four factors out of the six, namely:

a   The ground track.

b and **c**   The tidal stream, set and rate.

d   The vessel's intended speed through the water.

Correctly plotted, this will enable one to find:

e   The course.

f   The ground speed and thus the yacht's ETA.

The procedure will be as follows:

1   Plot the intended track, A to B.

2   Plot the tidal stream vector through the departure point A.

As can now be seen, the opportunity has arisen to draw one of two triangles, the correct and the incorrect! In fact, everything hinges on the direction of the tidal stream. In the diagram the flow is to the NNE and therefore the correct triangle lies to the N of the track line.

Rule 1: the correct triangle is always downstream of the track line.

Completing the course setting triangle
The temptation at this point is to join C and B. Do not, because (1) the length of AB is unlikely to represent your unknown ground speed, and (2) the length of CB is unlikely to represent your known water speed.

Rule 2: do not allow yourself to become obsessed with the destination, point B.

The correct, and only procedure, is to set your dividers at the yacht's intended water speed and, with centre C, find at which point the arc cuts AB or AB produced (extended).

Rule 3: do not find F by plotting the water speed along the track line AB.

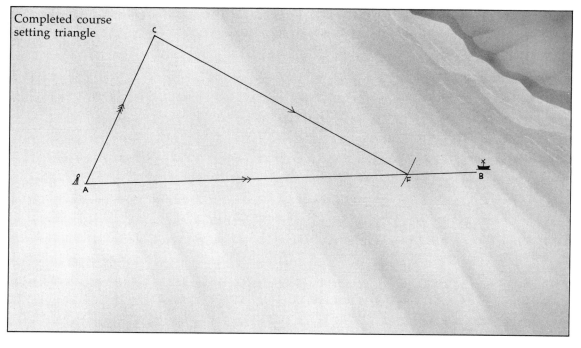

Completed course setting triangle

Note: point F may be short of, or beyond, B; or it could occasionally fall on it. Line CF represents the yacht's course and speed through the water. Let us now put this into practice.

Exercise 2

Chart 5043

a Find the correct course to steer from the East Goodwin Light Vessel to the North Goodwin Light Vessel. Speed 4.6 knots, tidal stream 044° at 2.2 knots. (keep the plots for **2a** and **2c** on the chart for use in Question 3.)

b Theoretically, on which side of the East Goodwin buoy would the vessel pass?

c Find the correct course to steer from the North Goodwin Light Vessel to Elbow buoy at 6 knots. Tidal stream 030° at 2.9 knots. *Note:* in this exercise it will be necessary to extend the track line.

We now need to tackle the question of estimated time of arrival or ETA. The best method is first to find the vessel's ground speed by scaling AF as shown in the previous diagram. By measuring the distance the vessel has to travel along her intended track AB, we can find the duration of the journey and hence the ETA, using the formula:

$$\text{Time (in minutes)} = \frac{\text{Distance (M)} \times 60}{\text{Ground speed (knots)}}$$

Exercise 3

Finding ground speed and ETA

a In exercise 2a and assuming a departure time of 0900 BST, find the ground speed and ETA.

b Find ground speed and ETA in Exercise 2c, using a departure time of 1021 BST.

The need for a half-hour triangle

Very often, especially when navigating a fast yacht and using a large scale chart, it is impossible to use a true triangle of velocities: there is simply not sufficient room on the chart. The solution is to halve, or occasionally quarter, all vector quantities. Care must be taken when carrying out such a plot to scale up the track line vector (AF) in order to get a true ground speed. Do not alter the scale on which *distances* are measured.

Exercise 4

Find the course and ETA at 9.0 knots from the NE Spit buoy (51°28′N, 1°30′E) sailing NNE to the Outer Tongue buoy. Tidal stream 298° at 1.6 knots. Departure 0800 BST.

Triangle of distances

Although it is not always necessary to construct this triangle, it is important that the navigator is aware of its existence and properties. It should serve to deepen his understanding of the principles involved when navigating in strong cross-tides in restricted visibility.

In this plot, the course line CF has been transferred to its proper position, originating at the departure point A. The tidal stream vector AC has been transferred to the destination point B. We now have a triangle of distances ABG. Point G would be the DR position for the time of arrival at B. The vector AG (the transferred course line) represents the amount of 'water' the boat has to sail through and is, therefore, another route to deriving ETA via *water speed*. The line GB is the actual drift experienced on the journey from A to B.

The plot

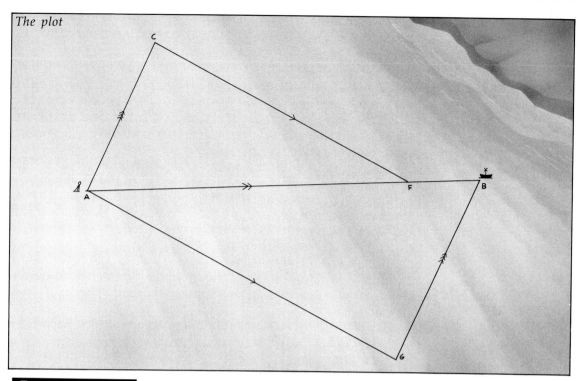

Exercise 5

Triangle of distances, chart 5043

a First find the course and ETA from CS2 buoy (51°08′.6N, 1°34′.0E) northwards to the East Goodwin Light Vessel at 6 knots, tidal stream 250° at 2.4 knots, by drawing a conventional triangle of velocities. Departure time 0900 BST.

b By transferring the course to the departure buoy CS2, and the tidal stream to the destination, the East Goodwin Light Vessel, construct the triangle of distances and find the length of the transferred course in cables.

Note: this confirms the duration of the journey as being about 67 min, i.e. 67 cables at 6 knots water speed.

c What was the drift in this exercise?

d Give the latitude and longitude of the DR at the time the East Goodwin Light Vessel was reached, i.e. DR at 1007 BST.

Exercise 6

Chart 5043

In the exercises that follow, short term course setting will be combined with tidal stream evaluation. It may be necessary be re-read the summary at the end of Chapter 4.

a Find the course from the Elbow buoy (51°23′.2N, 1°31′.7E) SSE to the North Goodwin Light Vessel. Use tidal diamond N. Speed is 4.2 knots, time of departure 1000 BST. HW Dover 1422 BST. Range 4.2 m.

b What is your ground speed and ETA?

c Give the yacht's DR position at the time of arrival at the Light Vessel.

Note: construct a triangle of distances.

a At 1310 BST your position is 51°25′.00N, 1°30′.00E. Find your course and ETA eastwards to Drill Stone buoy at 6.5 knots. HW Dover 1050 BST. Range 6.2 m. Use tidal diamond Q.

b Which tidal diamond may also have been relevant to Question 7a?

Moving towards course setting long term

In the previous question, the duration ran into the second hour (i.e. 71 min) and, had the trip been much longer, a second tidal stream vector would have been necessary. In Exercise 8 this situation also arises, only more so. At this stage, we will use the 'stay on track' method. To do this, the navigator simply constructs two triangles of velocities, the second being drawn at the point reached along the track line at the end of the first hour. Two courses will therefore arise, each designed to counteract the tidal stream and keep the yacht sailing along her designated track as closely as possible.

a Find the two courses needed and the ETA from the South Goodwin Light Vessel NE to the South Falls buoy at 5.8 knots. Departure time 2319 BST, HW Dover 2245 BST. Range 5.9 m (mean springs). Use a mean of tidal diamonds D and A for the first hour, and diamonds E and C for the second.

b Having reached the South Falls Buoy, what is the correct course (T) to steer so as to cross the SW-bound shipping lane?

c What track does this course produce, assuming a tidal stream of 025° at 2.7 knots and a water speed of 5.8 knots?

d How long will the vessel take to cross the SW-bound lane?

N.B. Traffic lanes should be crossed **with the boat heading at right angle to the direction of traffic in the lane**, regardless of tidal stream. This reduces the time taken to cross the lane to a minimum, and also means that the boat presents a clear aspect to lane users. The engine should be used if the speed under sail drops below about 3 knots.

Course setting, long term

The classic need for this procedure would be a situation such as the crossing of the English Channel, when a journey of several hours across the tidal stream in being contemplated. Having decided on a departure time and estimated the probable duration of the passage, the tidal stream atlas would be used to give the navigator a series of tidal vectors. This procedure was rehearsed in Chapter 4 (Example 5d). The next decision at this stage is whether or not to adopt the older 'stay on track' method or the newer ploy which is essentially a 'stay on course' method.

'Stay on track' v 'Stay on Course'

We saw the 'stay on track' method in use over a 2 hr period in the previous question (8a) and obviously it could be extended to deal with much longer journeys. Certainly it is the only method to use when dangers lie on either hand, necessitating the vessel to navigate along a strictly defined track. Such a situation is highly unusual: it certainly does not arise when crossing the English Channel, for example. In exercises on the Stanford chart, we may find dangers in our path such as Les Minquiers or Les Roches Douvres, but the obstructions simply turn the problem from a single leg passage into a two leg passage. The 'stay on track' method does not have to be used in such circumstances. By this stage, the reader will have gathered that the 'stay on course' method is the more favourable and should be practised whenever possible. Its advantages are as follows:

1 It is the quicker way of getting from A to B and so has enormous advantages when racing, or trying to catch the tide at your destination. This saving in time is particularly noticable when the passage encompasses a reversal of tidal stream.

2 Having only one course to steer throughout the voyage, it is much easier to administer and may have considerable advantages when the yacht is sailing just off the wind.

The only disadvantage is that it requires a little more time and skill to work out, especially if the navigator wants to see the exact (usually curved) track that will be followed.

It may be worth noting that the slavish use of an electronic navigator such as Decca results in the slower 'stay on track' method being followed. We shall see in Chapter 10, how a more imaginative use of an electronic navigator can make it possible to pursue the more efficient constant course method.

The best approach at this stage must be to show how the same passage can be accomplished using both methods.

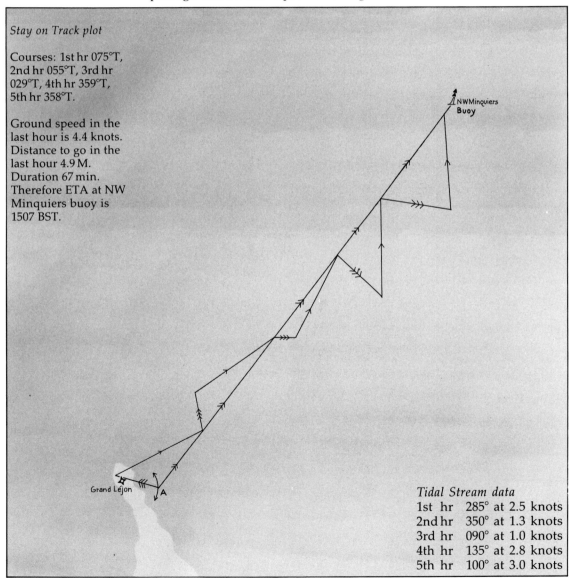

Stay on Track plot

Courses: 1st hr 075°T, 2nd hr 055°T, 3rd hr 029°T, 4th hr 359°T, 5th hr 358°T.

Ground speed in the last hour is 4.4 knots. Distance to go in the last hour 4.9 M. Duration 67 min. Therefore ETA at NW Minquiers buoy is 1507 BST.

NW Minquiers Buoy

Grand Léjon

A

Tidal Stream data

1st hr	285°	at 2.5 knots
2nd hr	350°	at 1.3 knots
3rd hr	090°	at 1.0 knots
4th hr	135°	at 2.8 knots
5th hr	100°	at 3.0 knots

Example *'stay on track' method on Stanford chart*
Calculate the courses and ETA from a position 1 M E of the Grand
Léjon Lighthouse (48°45′N, 2°40′W) to the NW Minquiers buoy (49°00
N′, 20′W) at 4 knots. The exercise commences at 1000 BST, distance
19 M, estimated duration 5 hr.

Exercise 9 Plot the data of the previous example on the Stanford chart.

'Stay on course' method

We shall now see how this can be carried out using the 'stay on
course' method. First plot the five tidal vectors in chronological sequence from our departure point, a mile to the E of the Grand Léjon
light. The reader should actually do this, so that the relevant measurements can be checked as we proceed. The next move requires a little
skill and patience: we must attempt to equate the duration of the
passage, with the tidal stream vectors. As we have seen with the
triangle of distances, duration can be determined by measuring along
the course line in terms of *water speed*. Your plot should appear as
follows:

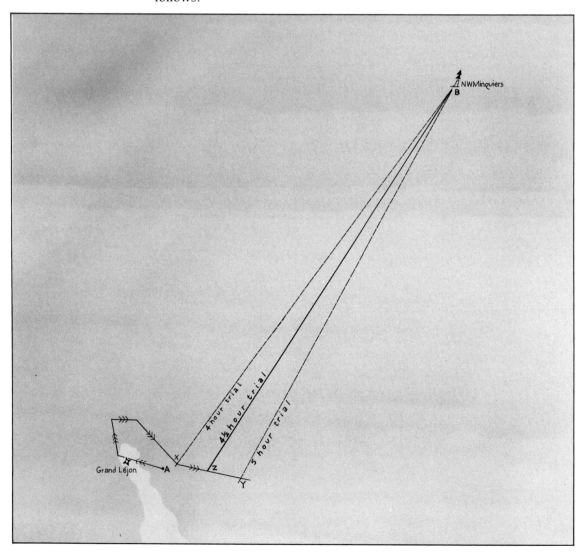

A is the point of departure , B the destination, X and Y are the end points of the fourth and fifth hours of tide respectively. We must now embark on a short series of trial measurements, based on estimations of the passage duration. First let us assume a 5 hr duration. Point Y represents 5 hr of tide and the distance YB should measure 20 M or 5 hr of sailing at 4 knots. However, YB is only 17.2 M. Therefore our equation does not balance and the plot must be less than 5 hr duration. Next let us assume a 4 hr duration. To check this, we must measure XB, hoping for 16 M. XB measures 18.5 M, so clearly the trip is longer than 4 hr. Point Z represents 4½ hr of tidal stream and the distance ZB is 18 M, or 4½ hr of sailing at 4 knots. The equation is now balanced and ZB is the course (033°T) and the ETA is 1430 hr. Compare this with the 'stay on track' ETA of 1507: about 37 min have been saved and only one course has to be steered throughout.

There is a third bonus for sailing craft. Should the wind be northerly (say 350°), then the latter courses of the 'stay on track' system would be unsteerable and an enormous amount of time would be expended in beating to windward. On a typical cross-Channel passage lasting one tidal stream cycle or more, very significant savings in time can be

made by avoiding the temptation to stay slavishly on track throughout It must be stressed, however, that when such a crossing is accomplished under sail alone, it is very unlikely that the original course will be kept continuously without amendment, due to inevitable changes in boat speed. The sailing navigator must therefore be prepared to review his course in the light of changes in wind strength or direction, when going across tidal streams of any significant strength.

Some readers may be concerned about the track of the vessel during the passage. It must be stressed that the track is not a straight line AB but is curved or sinuous. The straight line AB must *not* be drawn when using 'stay on course' method, as it is both irrelevant and misleading. Let us see how the vessel's track may be determined. First transfer the course line westwards from its starting point Z to the actual departure point A. Next clean the chart of all plot lines other than the transferred course line and the tidal vectors A to Y. Now plot a series of DR positions along the course line at 4 M intervals, up to and including no. 5 (20 M). Plot a series of EPs using the tidal vectors. The diagram that follows should make clear how this can best be achieved.

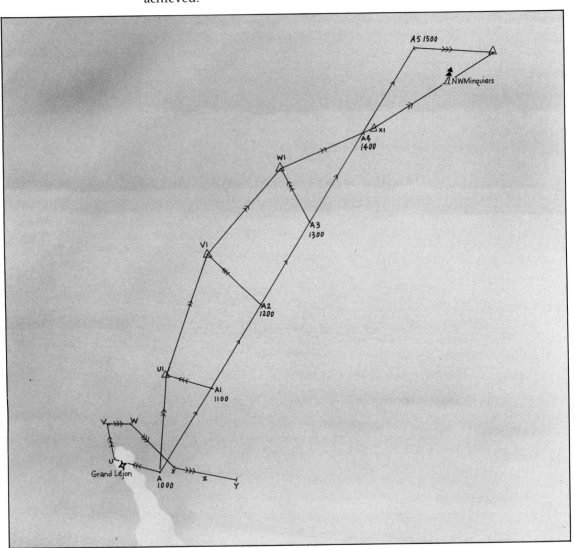

Example to plot the 1300 EP, transfer the vector AW from the vector plot at A to the 1300 DR at A_3. Draw the line A_3 W_1. *Note:* the NW Minquiers buoy B lies on the fifth hour track line X_1 Y_1. It would seem from this plot that the ETA at the NW Minquiers buoy will be slightly before 1430 BST.

Exercise 10

A practice in both 'stay on course' and 'stay on track' methods. Stanford chart. Guernsey to St Malo: first leg St Martin's Point to the SW Minquiers buoy. Speed 5 knots. Departure 0900 BST.

Tidal stream data:
1st hr 0900–1000 290° at 2.0 knots
2nd hr 1000–1100 270° at 2.5 knots
3rd hr 1100–1200 225° at 1.0 knots
4th hr 1200–1300 150° at 1.5 knots
5th hr 1300–1400 135° at 2.5 knots
6th hr 1400–1500 120° at 3.0 knots

a List the courses needed to stay on track for the SW Minquiers buoy.

b What is your ground speed during the hour 1300 to 1400?

c What is your ETA at the buoy using this method?

d What is your course for, and ETA at, the SW Minquiers buoy using the 'stay on course' method?

e Plot your track when using the 'stay on course' method. What is your EP at noon using this method?

f If the wind was westerly and the yacht was making 8° leeway, what would be her heading, magnetic?

g What is the heading on the compass? Use *Zephyr's* deviation card.

Exercise 11

Stay on course method
a Find the course and ETA from the NW Minquiers buoy to the Demie de Pas Light, 1½ M S of St Helier Harbour (beacon Morse **D**). Speed 4 knots, time of departure from NW Minquiers 1600 BST, date Friday, August 22.

Note: the following information will be needed:

1 Time (BST) and height of tide Dover for August 22.

2 Correct starting page in the *Stanford Tidal Atlas*.

3 An estimation of the maximum number of hours of tide that may be required.

4 An extract of the tidal streams from the atlas, page by page, hour by hour.
    All this information will be given with the answers to Exercise 11a.

b State the 1900 DR relative to La Corbiere Light (SW Jersey).

c State the 1900 EP relative to the Demie de Pas Light (Morse **D**).

Exercise 12

Stanford chart. Further practice in using the stay on course method.
a Find the course and ETA at the Bell buoy NNW of Cezembre (48°41'.0N, 2°05'.0W) from the SW Minquiers at 4 knots, given the following tidal streams:

1st hr   0800−0900   140° at 2.5 knots
2nd hr   0900−1000   120° at 2.0 knots
3rd hr   1000−1100   090° at 1.5 knots

Summary

Distance (miles) = Speed (knots) × Time (hours)

so $\text{Speed} = \dfrac{\text{Distance}}{\text{Time}}$

$\text{Time (minutes)} = \dfrac{\text{Distance (M)} \times 60}{\text{Speed (knots)}}$

Always associate ground speed with track, and water speed with course. The correct course setting triangle is always downstream of the track line.

Answers to Chapter 8

Exercise 1
Downs 0851 BST
South Brake 0902 BST
NW Goodwin 0917 BST
Gull Stream 0935 BST

Exercise 2
a   Course 328°T

Possible errors	Cause	Remedy
013°	Wrong triangle	Read Rule 1
321°	Water speed on track line	Read Rule 3
334°	Used destination point	Read Rule 2

b   Just to the W. Actual path is along the track line.
c   307°T

Exercise 3

a   Ground speed 5.5 knots. ETA 1021 BST $\left(\dfrac{7.14 \times 60}{5.5} \simeq 81 \right)$

b   Ground speed 7.0 knots. ETA = 1049 BST $\left(\dfrac{3.3 \times 60}{7.0} = 28\frac{1}{4} \right)$

Exercise 4
Course 023°. Ground speed 9.2 knots. ETA 0821 BST.

Exercise 5
a   Course 036°T. Ground speed 4.2 knots.

ETA 1007 BST $\left(\dfrac{4.7 \times 60}{4.2} \simeq 67.1 \text{ min} \right)$

b   Course line = 67 cables (6.7 M)
c   Drift is 2.65 M
d   DR position at 1007 BST 51°14'.0N, 1°40'.3E
Note: should you find that the answers to 5b, c and d all indicate a position too far south (i.e. for 5b, 60 cables or 5c, 2.4 M) then check that the tidal stream vector has been transferred to the East Goodwin Light Vessel and *not* the northern point of the triangle of velocities.

Exercise 6
a   126°T Tidal data: HW−4. Therefore 208°, 3.1 knots springs, 1.7 neaps. Calculated rate 2.2 knots.
b   5.0 knots, 1040 BST.
c   51°21'.55N, 01°35'.30E

Exercise 7
a   Course 096°T. Ground speed 6.9 knots. ETA 1421 BST. Tidal stream 016° at 1.4 knots.

b Tidal diamond P. A mean value of P and Q would have been preferable, say 035° at 1.4 knots.

Exercise 8
a 1st course 066°T. 2nd course 072°T. ETA 0040 BST.
Tidal streams:
1st hr diamond A 046°   3.0 knots mean 044°, 3.0 knots
1st hr diamond D 042°   3.0 knots
2nd hr diamond E 014°   3.1 knots mean by drawing 025°, 2.65 knots
2nd hr diamond C 040°   2.3 knots

b 130°T
c 103°T.
d 27½ minutes.

Exercise 9
Courses 068°T, 050°T, 024°T, 008°T, 356°T. ETA 1446 BST.

Exercise 10
a 1st hr   144°T   4th hr 169°T
    2nd hr 136°T   5th hr 179°T
    3rd hr 156°T   6th hr 190°T (not 199°T)
b 7.0 knots
c 1447 BST
d Course 161°T. ETA 1436 BST
e EP at noon is 49°10'.6N, 2°31'.7W. *Note:* this is the maximum cross-track error (see Chapter 10)
f Heading 175°M
g Heading 180°C

Exercise 11
HW Dover 1335 BST, 6.8 m. Starting page, 3 hr after HW Dover (1635 BST). Estimated duration less than 4 hr.

Page	Hour	Time	Set*	Drift*	
HW + 3	1st hr	1600–1700	135° at 1.4 knots		course 101°T
HW + 4	2nd hr	1700–1800	110° at 2.9 knots		
HW + 5	3rd hr	1800–1900	100° at 3.2 knots		ETA 1906 BST
HW + 6	4th hr	1900–2000	090° at 2.4 knots		

b Corbiere bears 120°, 1.4 M
c Light bears 013°, 4 cables
*Some discrepancies in the set and drift figures extracted from the atlas will be unavoidable, as such data is not entirely a matter of fact but is, to some extent, a matter of judgement.

Exercise 12
Course 157°T, ETA 1054 BST (2.9 hr)

Getting the Act Together

As the reader may now have realised, no one step in navigation is difficult. The main problem is to remember the steps in the correct sequence avoiding the pitfalls on the way. The requirement, at this stage, is for exercises that are both all-embracing and realistic, in terms of what is encountered in actual navigating. We need to get our act together.

Passage planning

Let us consider a passage from St Peter Port in Guernsey, southwest-wards to Lézardrieux in Brittany. Looking at the project purely from a navigational point of view, the first priority is to ensure that the correct reference literature is on board.

Exercise 1
Make a short list of (a) the mandatory publications and (b) the additional publications which it would be useful to have on board.

Exercise 2
Next we must consider the all important factors of timing and route. Study the tidal atlas and chart and (given that the forecast is for westerly winds) and decide if it is better to go E or W of the Roches Douvres, taking an assumed speed of 5 knots through the water into account.

Exercise 3
Assuming the date is Monday, August 18 and speed is 5 knots, what are your approximate departure and arrival times? Remember (from Answer 2) that you wish to be W of Roches Douvres at HW Dover + 3 hr, or approximately 1400 BST.

Exercise 4
Can you leave the inner marina in St Peter Port at 0900 BST? The marina sill is navigable for 2 hr either side of local HW.

The basic passage plan having been structured, the navigator would now need to spend time reading up the pilotage details for (a) leaving St Peter Port bound S, and (b) the approach to, and navigation of, the Trieux River. From the latter he would discover that the marina at Lézar-drieux is unusual, for Brittany, in that it is accessible throughout the tide. He would also need to obtain the latest weather forecast information.

Exercise 5
If an almanac is available, find the VHF channel numbers and times of the Channel Islands weather forecast service from Jersey Radio.

Exercise 6
Assuming that an average speed of 5 knots will be maintained, list the tidal streams that should be encountered between St Martin's Point, Guernsey, and the Sirlots buoy off Lézardrieux (48°53′.0N, 2°59′.5W). HW Dover 1056 BST, height 6.0 m. Departure from St Martin's Point is at 0930 BST.

Exercise 7
Using the tidal streams in the answer to Question 6 and employing the 'stay on course' method, find the course and ETA to a point 1½ M W of the Roches Douvres reefs, sailing at 5 knots.

Exercise 8
Find the course and ETA for the Sirlots buoy from the departure point 1½ M W of the Roches Douvres reefs. Time of departure 1315 BST.

The passage is now fully prepared, although the courses should be regarded as provisional as they depend on a speed of 5 knots being maintained throughout.

We will now assume that the cruise is actually undertaken and deal with some of the navigational eventualities as they arise, including a change in speed.

Log   *Variation 6°W*

0900 sailed. Log reading 0.0 M. Weather fair, wind W force 3 to 4. Forecast: sea slight, wind veering slowly and becoming force 2 to 4.

0927. Log 2.2 M. Off St Martin's Pt, sea moderate to rather rough due to weather-going tide and the strong winds of the previous day. Leeway estimated at 8°.

Exercise 9

Based on a course of 213°T, what heading (compass), should the helmsman be to told to steer?

The boat can just lay this heading and is jogging along at approximately 5 knots. Water speed indicator is oscillating due to confused sea.

1030 BST. Log 6.4 M. Hanois Light bears 332°M. St Martin's Pt bears 059°M. Right-hand edge of Little Sark bears 085°M.

Exercise 10

Plot these three bearings and explain why:

a The bearings do not intersect perfectly.

b The fix shows a discrepancy when compared with the 1030 EP provisionally plotted before departure.

Exercise 11

From the centre of the 1030 fix, what is the new course based on a speed of 4.0 knots and the tidal stream data in Answer 6?

The cruise continues at a slower pace, the sea gradually becoming less rough and the wind veering but becoming lighter.

1130 BST. Log 10.5 M. Visibility about 6 to 8 M. Decided to do a running fix on Roches Douvres Light as soon as it is clearly visible.

1220 BST. Roches Douvres sighted.

Exercise 12

1225 BST. Log 14.4 M. Roches Douvres bears 193°M. Course 212°T.

1310 BST. Log 17.5 M. Roches Douvres bears 146°M.

Plot the 1310 running fix position and state the distance off the light structure.

Exercise 13

By projecting your track, find out if you are on course for your turning point, 1.5 M W of Roches Douvres reefs.

Exercise 14

1358 hours. Log 20.5 M. The following bearings are obtained: Roches Douvres Light 081°M, Barnouic light structure 150°M. Plot this position and re-calculate the course (T) for Les Sirlots buoy at 4 knots based on the following tidal stream data:

1400−1430 BST   100° at 1.2 knots (plot 0.6 M)
1430−1530 BST   100° at 2.7 knots
1530−1630 BST   090° at 2.4 knots
1630−1730 BST   090° at 2.4 knots

Exercise 15

Re-calculate the course and ETA on a motor-sailing speed of 6 knots.

Exercise 16

Plot your track between 1430 and 1530 BST to see if it clears the Plateau de Barnouic, based on the fix at 1355, a speed of 6 knots and the tidal stream data in Question 14.

Exercise 17

Calculate sea level for the Isle de Bréhât (No.1621) for 1630, in order to assess the danger of the reefs NNW of the Sirlots buoy; namely Carrec Mingue (11 ft sounding) and Carrec Dône (less than 6 ft sounding: symbol is +).

Exercise 18

At 1600 BST, log 32.7 M, the following bearings are obtained: Les Héaux Lighthouse 266°M. La Horaine Lighthouse 156°M. Plot your position and determine if you are E or W of track.

Exercise 19

a What alteration of course is needed?

b What is the new course?

c What is the new compass heading? Leeway zero as the yacht is now motoring into negligible wind.

Exercise 20

Give a possible reason why the yacht was E of track at 1600.

2100 FST. Log zero. Position La Vielle Tour Restaurant, Paimpol.

The last two exercises in this chapter test your ability to use the triangle of velocities to solve: (a) the problem of turning back when overtaken by fog in the initial stages of a passage, and (b) to find the optimum time to tack when beating to windward in a strong tide.

Exercise 21

Out and Return. Chart 5043.
Time 0830 BST. Log 1.5 M. Position between outer buoys of the dredged channel leading into Ramsgate Harbour (51°19′.50N, 1°27′.35E). Course 065°T. At 0915, log 5.8 M, a decision is taken to return to port due to fog. Calculate the return course and ETA at 5.2 knots. Tidal streams are: outward leg 015° at 2.4 knots, return leg 345° at 2.0 knots.

Exercise 22

When to tack
Chart 5043. Variation 5°W. *Note:* this is a racing situation in which one of the marks is the North Goodwin Light Vessel.

The yacht is beating to windward, closehauled on port tack, heading 060°M, leeway 8°. Her present position is in the vicinity of the tidal diamond L (51°18′.2N, 1°34′.4E). Speed 5.4 knots. The navigator knows:

a The vessel appears to tack through 90° (heading to heading).

b The tidal stream is 007° at 2.4 knots.

He wishes to calculate the bearing (mag) of the North Goodwin Light Vessel, when the optimum position to tack has been reached. *Hint:* concentrate on the triangle of velocities appropriate to the new starboard tack.

Pilotage

No book on coastal navigation would be complete without at least a mention of pilotage — the technique of navigating close to land and in the approaches to harbours by reference to marks (either natural or man-made) which can be identified visually and on the chart. Although in previous chapters we have stressed the importance of knowing the vessel's latitude and longitude, in pilotage waters this becomes secondary to her position relative to one or more predetermined marks.

In pilotage waters situations can develop very quickly, and so it is important to study the Sailing Directions carefully well before entering a strange harbour. There may, for example, be traffic signals which control entry and departure. Find out what the principal navigation aids and other marks consist of (there may be illustrations in the Pilot books) and where they are located on the chart. Having worked out the height of tide on arrival, identify any shoals that must be avoided, and decide where you intend to moor or anchor.

Certain bearings of objects may indicate safe water (clearing lines) — or just the opposite (danger bearings). Leading lines or transits (with two objects in line) should be used whenever they are available, since they indicate the best water. But to avoid any possible error always note the compass course when steering down a transit, and compare it with what it should be from the chart or Sailing Directions. Such bearings should be recorded beforehand, since in a tricky channel there is often little or no time for working at the chart table — indeed it may be necessary to refer to the chart in the cockpit.

Exercise 1

a Passage charts. Stanford chart of the Channel Islands or Admiralty chart BA 2668. Almanac. *North Brittany Pilot*, *Channel Islands Pilot*. Channel Islands Tidal Stream Atlas. Passports and Ship's papers in order.

b Current *Small Craft Edition of Notices to Mariners*. Guernsey large scale chart BA 807 or 808. Large scale chart of Lézardrieux, either French or Admiralty.

Exercise 2

Although E about Roches Douvres is the more direct route, it would be better to sail W about because of the tidal streams. This would also avoid the area of magnetic anomaly. The optimum passage plan would put the yacht W of Roches Douvres just before the flood starts (3 hr after HW Dover), thus putting her slightly up-wind and up-tide of her destination.

Exercise 3

Depart St Peter Port at 0900 BST. Arrive Lézardrieux Marina at approximately 1800 BST. 9 hr passage. HW Dover approximately 1100.

Exercise 4

No. Local HW Jersey is 0611 BST. HW Guernsey will be at about the same time or slightly later. Must leave by 0800 and wait for 1 hr in the outer harbour.

Exercise 5

Channels 25 or 82. 0645, 1245, 1845, 2245 GMT.

Exercise 6

1st hr	0930–1030 BST	270° at 1.0 knots	
2nd hr	1030–1030 BST	225° at 2.0 knots	
3rd hr	1130–1230 BST	225° at 1.6 knots	
4th hr	1230–1330 BST	240° at 0.7 knots	average of 1.0 and 0.4 knots
5th hr	1330–1430 BST	100° at 1.4 knots	average of 1.6 and 1.2 knots
6th hr	1430–1530 BST	100° at 2.4 knots	
7th hr	1530–1630 BST	090° at 2.4 knots	

Exercise 7

Course 213°T. Duration 3.75 hr. ETA 1315 BST.

Exercise 8

Course 218°T. Duration 3 hr 6 min. ETA 1621 BST. *Note:* course difference is 5° to *starboard* for Lézardrieux due to tide. This passage could have been treated as having only one leg rather than two, and proves the point that W about Roches Douvres is the better course to take.

Exercise 9

229°C.	Course 213°T.
	Heading 221°T.
	Heading 227°M. Heading 229°C

Exercise 10

a Sea being moderate to rather rough, precise bearings are therefore difficult to obtain.

b Ship sailing slower than predicted. Tidals stream stronger than predicted or leeway allowance excessive.

Exercise 11

Course 212°T. *Note:* the 1030 fix to the W of the original track has cancelled the effect of the E-going tide in the last hour. New course is substantially the same as the original.

Exercise 12

2.4 M off Roches Douvres (tidal stream 240° at 0.7 knots).

Exercise 13

Yes, but the young flood tide is predicted to commence at 1330 (100° at 1.4 knots). Should therefore alter course to starboard (luff up) by 10°.

Exercise 14

Situation is impossible at 4 knots. Engine must be started if Lézardrieux is to be reached in a reasonable time.

Exercise 15

Course 223°T. ETA 1645.

Exercise 16

Well clear. Should pass 1.8 miles W of Whistle buoy.

Exercise 17

Sea level equals 8.4 m approximately, therefore reefs present no danger. (HW Bréhat 1806 BST 9.8 m, LW 1.6 m. Time 1 hr 36 min before HW)

Exercise 18

East (about 0.6 M)

Exercise 19

About 7° to starboard. Course 230°T. Heading 238°C.

Exercise 20

Tidal stream stronger that predicted in the approaches to the river.

Exercise 21

Return course 210°T. ETA 1037 BST (ground speed 4.1 knots)

   Method: plot EP for 0915, based on a logged distance of 4.3 M and a drift of 1.8 M (¾ of 2.4 M).

   From the 0915 EP, carry out a short term course setting exercise, using a full hour triangle of velocities, designed to return the yacht along its original track.

Exercise 22

Tack when Light Vessel bears 337°M.

Method: calculate the true *course* when on starboard tack thus:

060°M	Heading port tack
330°M	Heading starboard tack
325°T	Heading starboard tack
317°T	Course starboard tack

Now draw a triangle of velocities (this can be done anywhere on the chart) to find the yacht's true track (332°T). Lastly, express the track angle magnetic (337°M).

Navigating with the Aid of Decca

The overwhelming factor that leads to the choice of a Decca-type navigator is continuity of information. The existing forms of satellite navigator, while they are invaluable to the ocean-going yachtsman, have the great disadvantage of giving a position update at 1 to 2 hour intervals. For most of us, the Decca coverage from Gibralter to Norway, including the whole of the British Isles and the Baltic, is sufficient for our cruising needs. Coupled with an update every 20 seconds and high 24 hour accuracy with continuous access, the choice is easily made. Some satellite navigators can be interfaced with the ship's log and compass, thus providing a continuous 'estimated position' between satellite fixes, but this adds greatly to the cost and still relies on the navigator to key in his assessment of tidal streams and any currents.

Having decided in favour of Decca and raised the necessary funds for equipment and installation, the next choice lies between the various makes. Some are Decca approved and some are not. There are few arguments in favour of buying a non-approved set. For many yachtsmen, the Decca Mark III will be the obvious choice. Although this set is marginally less accurate at extreme ranges than some of the more expensive receivers, it has a very high order of accuracy in most of the popular cruising areas. The exception is the Bay of Biscay, where Decca coverage is from rather remote stations and accuracy is of a lower order than normal. Under most conditions, a fix to within 1.3 cables (0.13 M) is to be expected, and the set will often give fixes accurate to the limit of its readability, 0.1 cable (0.01 M). This does mean that in poor visibility Decca can be used as an aid to pilotage, especially if it has been possible to calibrate the instrument in that area (see below).

Installation

Obviously the main source of guidance and advice must be from the manufacturers, but some additional comments might be useful.

Siting the receiving aerial The masthead is the first choice, but should this be too difficult the stern rail is almost as good. Decca signals are essentially ground waves and a high aerial is not really very important in most areas. Should you choose a masthead installation, then ensure that the aerial is at least half a metre away from the transmitting aerial of the yacht's VHF radiotelephone. Wherever the Decca aerial is installed, a low-impedance connection to earth must be made through a suitable submerged metallic skin fitting.

Noise suppression Like all radio aids to navigation, a Decca set will benefit greatly from an atmosphere that is interference-free. On some sets it is possible to monitor interference or 'noise' levels, and thus sources within the yacht herself can be identified. Possible sources that can easily be switched off should be checked first; this could include pumps, refrigerators and fluorescent lights. More difficult to deal with is the engine alternator, which will generate a great deal of interference, especially when delivering a large amount of current. The best way to deal with this problem is to fit a switch to its field coil circuit and thus turn off the alternator when using the Decca set in a critical situation. Some form of reminder that this has been done is necessary or flat batteries will result. Please note that it is most inadvisable to interrupt the flow of current from the alternator to the batteries: it will damage the alternator. A severe form of interference can arise from the rotating propeller. This can be very effectively suppressed by fitting brushes to the propeller shaft and connecting them to the ship's sacrificial anode. Such brushes are made by M.G. Duff of Chichester and marketed under the name Electro-Eliminator. They have the additional advantage of reducing propeller corrosion, which is the purpose for which they were designed.

Calibration In some areas the Decca system has a small distinct error that is wholly repeatable. It is thus possible to record the position displayed on the receiver as one passes close to a beacon or pierhead and compare this with the charted position. An 'offset' can then be entered which will automatically correct the error. The offset must be removed manually when the area has been left. However, it can be logged and re-entered on returning. Such a precaution is well rewarded when the fog comes rolling in.

Intelligent use of Decca

Although this system, and the even more superior systems that will surely follow, have revolutionised the practicalities of navigation, it is essential that the navigator uses Decca with discretion and does not fall into the trap of following it blindly. It must always be remembered that such systems are merely *aids to navigation* and that the overall task still has to be carried out by the navigator himself. It is essential that the quality of the information displayed is continuously assessed, especially when the navigational situation looks like being critical.

Let us take the example of leaving a river anchorage in poor visibility, with several twists and turns to negotiate before open sea is reached. Some would say 'don't go', others 'we have Decca, therefore there is no problem'. As is usual, the best course of action lies between the two extremes. In such a situation, the following assessment should be made:

1 Is the tide rising or falling? With the possibility of going aground, a rising tide is much to be preferred.

2 Assess the potential accuracy of the set. Note the maximum displayed error and the current noise levels.

3 How does Decca perform in this area? How does the displayed position check against an observed position? In other words, should an offset be entered?

4 What are the maximum permitted errors in the channels through the river estuary? How do these compare with the maximum displayed error?

5 Are there thunderstorms forecast which might temporarily render the set useless?

There are other parameters, but these are the main ones as far as Decca assisted pilotage is concerned.

Checking waypoints

All Decca receivers allow the user to enter a complete passage into the set, each turning point being known as a waypoint. It is, therefore, essential that each waypoint is checked against possible human error after entry. The most satisfactory method of carrying out such a check is to measure the track and distance between each point from the chart, at the same time as the latitudes and longitudes are determined. The same information can then be called up on the display. A good correlation must be obtained.

Using Decca to assist the navigator to make a 'stay on course' passage

It has already been said that the Decca system encourages a 'stay on track' approach to passage making, and the disadvantages of such an approach have also been made clear. With correct passage planning, however, it is perfectly possible to use a Decca set to aid a 'stay on course' method.

The best procedure is to plot the tidal streams as already described, and should they tend to cancel out then a 'stay on course' strategy will have advantages. Having completed the plot, put in the rhumb line from departure point A to destination B and measure the distances, at right angles, from the rhumb line AB to the various estimated positions at hourly intervals. This will give the intended departures from the rhumb line, which will be displayed on most sets as 'cross-track errors'. In this way the passage will be assisted and monitored by the Decca Navigator. Alternatively, the maximum cross-track error can be determined and put in as an extra waypoint.

This system is best understood by carrying out the following exercise.

Exercise 1

Decca assisted 'stay on course' passage
List the cross-track errors, and find the course and ETA in the following passage.

Departure point 1.0 M E of Grand Lejon Light 48°45'.0N, 2°40'.0W. Time 0900 BST. Speed 5.0 knots. Landfall St Martin's Point, Guernsey. *Tidal streams:*

0900–1000	100° at 2.5 knots
1000–1100	080° at 2.0 knots
1100–1200	050° at 1.5 knots
1200–1300	020° at 1.5 knots
1300–1400	000° at 1.5 knots
1400–1500	330° at 2.0 knots
1500–1600	300° at 2.5 knots

Answers to Chapter 10

Exercise 1

Course 002°T	ETA 1545 BST

Cross-track errors, measured from the rhumb line AB

1st hr	1000	2.2 miles E of rhumb line
2nd hr	1100	3.6 miles E of rhumb line
3rd hr	1200	4.2 miles E of rhumb line
4th hr	1300	4.1 miles E of rhumb line
5th hr	1400	3.7 miles E of rhumb line
6th hr	1500	2.2 miles E of rhumb line

Appendices

DOVER

MEAN SPRING AND NEAP CURVES
Springs occurs 2 days after New and Full Moon.

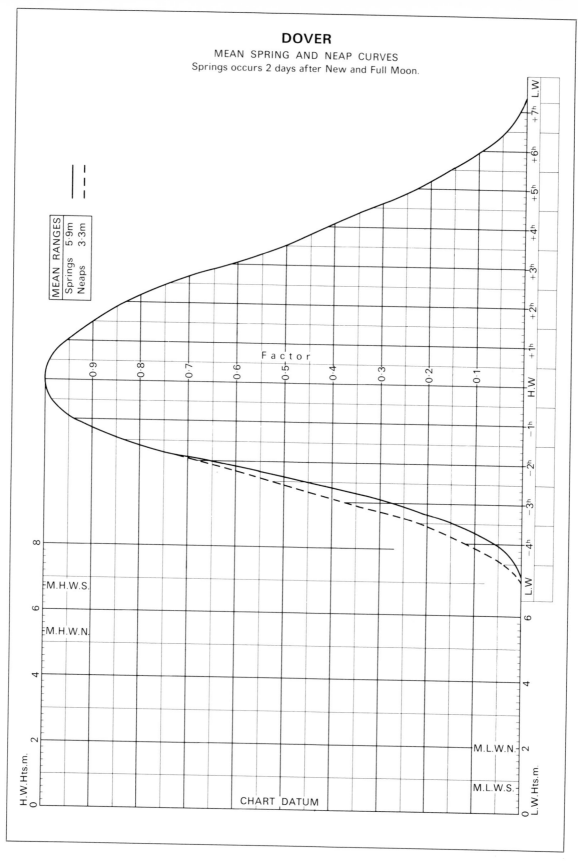

ENGLAND, SOUTH COAST - DOVER

LAT 51°07′N   LONG 1°19′E

TIMES AND HEIGHTS OF HIGH AND LOW WATERS

TIME ZONE GMT

YEAR 1986

MAY

Day	Time	M	Time	M	Time	M	Time	M
1 TH	0404	5.5	1130	1.8	1633	5.4		
2 F	0005	1.7	0532	5.2	1250	1.9	1801	5.2
3 SA	0130	1.6	0716	5.3	1412	1.7	1933	5.4
4 SU	0249	1.4	0820	5.6	1521	1.4	2030	5.7
5 M	0353	1.1	0907	5.8	1620	1.2	2112	6.0
6 TU	0447	0.9	0942	6.0	1708	1.1	2149	6.2
7 W	0530	0.9	1014	6.1	1746	1.1	2226	6.3
8 TH	0605	1.0	1048	6.2	1812	1.1	●2301	6.4
9 F	0632	1.0	1125	6.3	1836	1.1	2336	6.4
10 SA	0656	1.0	1200	6.2	1904	1.0		
11 SU	0008	6.3	0726	1.0	1231	6.2	1938	1.1
12 M	0034	6.1	0758	1.0	1259	6.0	2013	1.2
13 TU	0057	6.0	0832	1.2	1326	5.9	2049	1.4
14 W	0126	5.8	0905	1.4	1359	5.7	2127	1.6
15 TH	0205	5.5	0943	1.7	1444	5.5	2207	1.8
16 F	0258	5.3	1027	1.9	1545	5.3	2259	1.9
17 SA	0419	5.1	1123	2.1	1704	5.3		
18 SU	0004	1.9	0550	5.1	1236	2.1	1819	5.4
19 M	0124	1.7	0656	5.4	1355	1.9	1920	5.7
20 TU	0234	1.4	0748	5.7	1500	1.6	2011	6.1
21 W	0334	1.1	0836	6.1	1556	1.3	2057	6.4
22 TH	0426	0.9	0921	6.3	1647	1.1	2142	6.6
23 F	0513	0.8	1007	6.5	1734	0.9	○2228	6.7
24 SA	0601	0.7	1057	6.6	1822	0.8	2318	6.7
25 SU	0649	0.7	1149	6.6	1912	0.8		
26 M	0010	6.6	0740	0.7	1242	6.5	2002	0.8
27 TU	0104	6.4	0832	0.9	1333	6.3	2054	0.9
28 W	0159	6.2	0927	1.1	1420	6.1	2150	1.1
29 TH	0253	5.9	1024	1.4	1512	5.9	2251	1.3
30 F	0352	5.6	1125	1.6	1610	5.7	2354	1.5
31 SA	0502	5.4	1228	1.7	1719	5.5		

JUNE

Day	Time	M	Time	M	Time	M	Time	M
1 SU	0102	1.5	0625	5.4	1333	1.8	1836	5.6
2 M	0212	1.5	0730	5.5	1439	1.7	1941	5.7
3 TU	0315	1.4	0820	5.6	1536	1.6	2032	5.9
4 W	0409	1.3	0905	5.9	1624	1.5	2117	6.0
5 TH	0451	1.3	0946	5.9	1701	1.5	2159	6.1
6 F	0525	1.3	1026	6.1	1733	1.4	2238	6.2
7 SA	0554	1.2	1105	6.2	1805	1.3	●2316	6.2
8 SU	0627	1.1	1142	6.2	1842	1.2	2350	6.1
9 M	0703	1.1	1214	6.2	1920	1.2		
10 TU	0018	6.0	0738	1.1	1245	6.1	1958	1.2
11 W	0046	6.0	0813	1.2	1316	6.1	2034	1.3
12 TH	0120	5.9	0849	1.4	1351	6.0	2111	1.4
13 F	0201	5.7	0925	1.5	1433	5.8	2150	1.5
14 SA	0251	5.6	1004	1.7	1524	5.7	2235	1.6
15 SU	0355	5.5	1052	1.8	1626	5.7	2329	1.7
16 M	0504	5.4	1150	1.9	1730	5.7		
17 TU	0034	1.6	0608	5.5	1259	1.9	1832	5.8
18 W	0142	1.5	0707	5.7	1409	1.7	1930	6.0
19 TH	0247	1.4	0857	5.9	1514	1.6	2025	6.2
20 F	0348	1.2	0857	6.1	1614	1.4	2118	6.3
21 SA	0447	1.1	0953	6.3	1715	1.2	2213	6.4
22 SU	0546	1.0	1049	6.4	1812	1.0	○2309	6.5
23 M	0645	0.9	1144	6.5	1910	0.9		
24 TU	0007	6.5	0742	0.9	1235	6.5	2005	0.9
25 W	0059	6.4	0836	0.9	1320	6.4	2057	0.8
26 TH	0148	6.3	0927	1.0	1404	6.3	2148	0.9
27 F	0236	6.1	1014	1.2	1449	6.2	2237	1.1
28 SA	0325	5.8	1101	1.5	1539	6.0	2326	1.3
29 SU	0419	5.6	1147	1.7	1634	5.8		
30 M	0017	1.6	0522	5.5	1236	1.9	1739	5.7

JULY

Day	Time	M	Time	M	Time	M	Time	M
1 TU	0114	1.8	0631	5.4	1333	2.1	1849	5.6
2 W	0220	1.9	0735	5.4	1437	2.1	1954	5.7
3 TH	0321	1.9	0832	5.6	1536	2.0	2049	5.7
4 F	0409	1.8	0922	5.7	1623	1.8	2138	5.8
5 SA	0449	1.6	1006	5.9	1704	1.6	2221	5.9
6 SU	0527	1.5	1045	6.1	1744	1.4	2259	6.0
7 M	0607	1.3	1122	6.2	1827	1.3	●2332	6.0
8 TU	0646	1.2	1154	6.2	1906	1.2		
9 W	0001	6.1	0724	1.2	1225	6.3	1945	1.1
10 TH	0032	6.1	0801	1.2	1259	6.3	2020	1.2
11 F	0109	6.1	0832	1.3	1335	6.2	2054	1.2
12 SA	0148	6.0	0904	1.3	1413	6.2	2129	1.2
13 SU	0232	5.9	0941	1.5	1457	6.1	2209	1.3
14 M	0322	5.6	1021	1.6	1548	6.0	2254	1.5
15 TU	0420	5.7	1109	1.8	1645	5.9	2347	1.6
16 W	0522	5.6	1210	1.9	1747	5.8		
17 TH	0055	1.7	0627	5.6	1324	2.0	1853	5.8
18 F	0209	1.7	0735	5.6	1443	1.9	2002	5.8
19 SA	0322	1.6	0847	5.8	1559	1.6	2110	6.0
20 SU	0435	1.4	0953	6.1	1709	1.3	2214	6.2
21 M	0544	1.2	1049	6.3	1812	1.1	○2311	6.4
22 TU	0649	1.0	1136	6.5	1912	0.8		
23 W	0000	6.5	0745	0.9	1219	6.6	2004	0.7
24 TH	0046	6.5	0833	0.8	1300	6.6	2050	0.6
25 F	0127	6.4	0914	0.9	1340	6.6	2132	0.8
26 SA	0208	6.2	0949	1.2	1420	6.4	2209	1.0
27 SU	0250	6.0	0952	1.4	1504	6.3	2244	1.3
28 M	0336	5.8	1049	1.7	1552	6.0	2319	1.7
29 TU	0428	5.5	1123	2.0	1647	5.7		
30 W	0000	2.0	0534	5.3	1211	2.3	1757	5.4
31 TH	0056	2.2	0655	5.2	1316	2.4	1919	5.3

AUGUST

Day	Time	M	Time	M	Time	M	Time	M
1 F	0213	2.3	0805	5.3	1444	2.4	2026	5.4
2 SA	0328	2.2	0901	5.5	1552	2.1	2121	5.5
3 SU	0420	1.9	0946	5.8	1642	1.7	2204	5.7
4 M	0505	1.6	1024	6.0	1727	1.4	2238	5.9
5 TU	0550	1.4	1057	6.2	1811	1.2	●2306	6.0
6 W	0632	1.2	1127	6.4	1853	1.1	2336	6.2
7 TH	0712	1.2	1200	6.5	1931	1.0		
8 F	0010	6.3	0745	1.1	1234	6.5	2004	1.0
9 SA	0046	6.4	0812	1.1	1310	6.5	2033	1.0
10 SU	0124	6.3	0840	1.2	1347	6.4	2104	1.0
11 M	0204	6.2	0914	1.3	1425	6.3	2141	1.2
12 TU	0247	6.0	0952	1.5	1508	6.1	2221	1.4
13 W	0339	5.8	1037	1.7	1604	5.9	2313	1.7
14 TH	0441	5.6	1134	2.0	1711	5.6		
15 F	0022	1.9	0556	5.4	1257	2.2	1831	5.4
16 SA	0149	2.0	0726	5.4	1432	2.0	2002	5.5
17 SU	0319	1.8	0857	5.7	1559	1.7	2124	5.9
18 M	0438	1.5	0956	6.0	1709	1.3	2220	6.1
19 TU	0549	1.2	1042	6.4	1812	0.9	○2306	6.4
20 W	0648	0.9	1122	6.6	1906	0.7	2346	6.5
21 TH	0735	0.8	1158	6.7	1952	0.6		
22 F	0022	6.6	0815	0.8	1235	6.8	2030	0.6
23 SA	0059	6.5	0846	0.9	1312	6.7	2103	0.8
24 SU	0135	6.3	0910	1.2	1348	6.6	2129	1.0
25 M	0212	6.1	0929	1.4	1426	6.4	2155	1.4
26 TU	0251	5.9	0955	1.7	1505	6.0	2224	1.7
27 W	0336	5.5	1030	2.0	1550	5.6	2302	2.1
28 TH	0433	5.2	1118	2.4	1654	5.1	2354	2.4
29 F	0605	4.9	1219	2.6	1842	4.9		
30 SA	0107	2.6	0734	5.0	1351	2.6	2005	5.0
31 SU	0244	2.4	0834	5.3	1521	2.3	2101	5.3

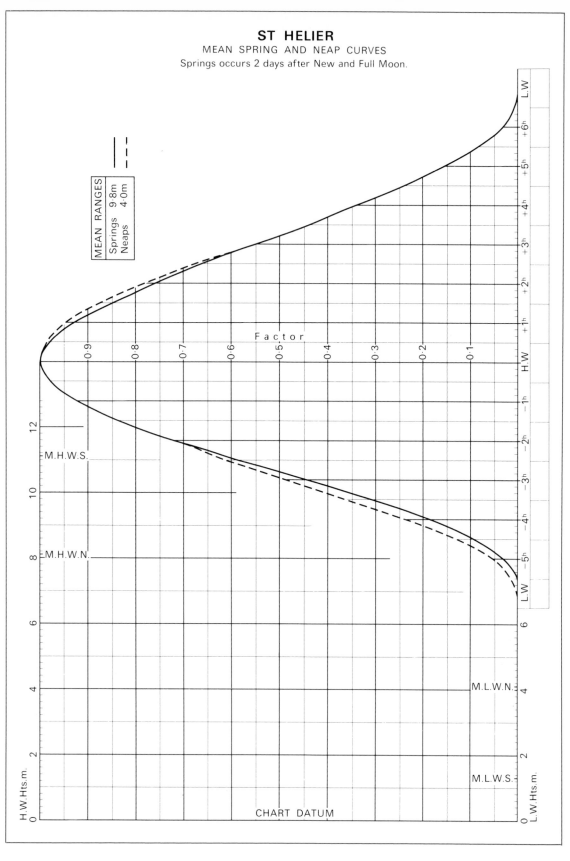

ST HELIER
MEAN SPRING AND NEAP CURVES
Springs occurs 2 days after New and Full Moon.

MEAN RANGES
Springs 9·8m
Neaps 4·0m

Factor

0·9 0·8 0·7 0·6 0·5 0·4 0·3 0·2 0·1

M.H.W.S.

M.H.W.N.

M.L.W.N.

M.L.W.S.

CHART DATUM

H.W.Hts.m.

L.W.Hts.m.

L.W +6ʰ +5ʰ +4ʰ +3ʰ +2ʰ +1ʰ H.W -1ʰ -2ʰ -3ʰ -4ʰ -5ʰ L.W

CHANNEL ISLANDS - ST. HELIER

LAT 49°11'N LONG 2°07'W

TIME ZONE GMT

TIMES AND HEIGHTS OF HIGH AND LOW WATERS

YEAR 1986

MAY

Day	Time/M	Time/M	Time/M	Time/M		Day	Time/M	Time/M	Time/M	Time/M
1 TH	0551 3.7	1144 8.1	1831 4.4			16 F	0454 4.0	1045 7.9	1716 4.3	2313 8.1
2 F	0031 8.0	0721 4.1	1334 7.8	2015 4.5		17 SA	0554 4.3	1208 7.7	1829 4.5	
3 SA	0211 8.2	0857 3.8	1457 8.2	2135 3.9		18 SU	0042 8.0	0717 4.2	1337 8.0	1958 4.2
4 SU	0317 8.7	1003 3.2	1552 8.8	2228 3.3		19 M	0202 8.4	0837 3.7	1443 8.7	2108 3.6
5 M	0407 9.3	1051 2.6	1634 9.4	2311 2.7		20 TU	0303 9.1	0939 3.0	1536 9.5	2204 2.8
6 TU	0449 9.8	1130 2.2	1711 9.9	2347 2.2		21 W	0355 9.9	1034 2.3	1624 10.3	2257 2.0
7 W	0527 10.2	1207 1.9	1746 10.2			22 TH	0444 10.6	1125 1.7	1712 10.9	2349 1.4
8 TH	0022 1.9	0603 10.3	1241 1.8	1818 10.4 ●		23 F	0533 11.1	1215 1.2	1800 11.3 ○	
9 F	0056 1.8	0636 10.4	1313 1.8	1850 10.4		24 SA	0039 1.1	0621 11.3	1307 1.1	1846 11.5
10 SA	0128 1.8	0710 10.3	1344 1.9	1921 10.4		25 SU	0131 1.0	0710 11.3	1357 1.1	1933 11.3
11 SU	0201 1.9	0741 10.1	1415 2.2	1952 10.2		26 M	0222 1.1	0757 11.0	1446 1.5	2019 10.9
12 M	0232 2.2	0812 9.8	1444 2.5	2022 9.9		27 TU	0310 1.5	0844 10.4	1532 2.1	2105 10.3
13 TU	0304 2.6	0843 9.4	1515 2.9	2053 9.5		28 W	0357 2.0	0934 9.8	1619 2.8	2156 9.6
14 W	0335 3.0	0914 8.9	1548 3.4	2127 9.0		29 TH	0448 2.7	1028 9.0	1711 3.5	2254 8.9
15 TH	0410 3.5	0952 8.4	1624 3.9	2210 8.5		30 F	0543 3.3	1133 8.4	1812 4.0	
						31 SA	0004 8.5	0652 3.7	1249 8.1	1930 4.2

JUNE

Day	Time/M	Time/M	Time/M	Time/M		Day	Time/M	Time/M	Time/M	Time/M
1 SU	0120 8.4	0806 3.7	1402 8.2	2042 4.0		16 M	0628 3.7	1238 8.5	1904 3.8	
2 M	0226 8.6	0911 3.5	1500 8.5	2138 3.6		17 TU	0103 8.7	0741 3.5	1348 8.8	2018 3.5
3 TU	0319 8.9	1002 3.1	1548 8.9	2223 3.1		18 W	0212 9.1	0851 3.1	1453 9.3	2124 2.9
4 W	0407 9.2	1044 2.8	1630 9.3	2304 2.7		19 TH	0317 9.6	0956 2.6	1552 9.9	2224 2.4
5 TH	0451 9.5	1123 2.6	1709 9.6	2344 2.5		20 F	0416 10.1	1057 2.1	1648 10.4	2325 1.9
6 F	0532 9.7	1203 2.4	1749 9.9			21 SA	0513 10.5	1156 1.8	1743 10.8	
7 SA	0024 2.3	0612 9.8	1241 2.3	1827 10.0 ●		22 SU	0024 1.5	0610 10.7	1253 1.5	1835 11.0 ○
8 SU	0104 2.2	0703 9.8	1320 2.4	1902 10.0		23 M	0123 1.2	0703 10.8	1349 1.5	1927 11.0
9 M	0144 2.2	0727 9.7	1357 2.4	1938 9.9		24 TU	0216 1.2	0755 10.7	1440 1.6	2015 10.9
10 TU	0220 2.3	0802 9.5	1433 2.6	2012 9.8		25 W	0305 1.3	0843 10.5	1527 1.9	2101 10.5
11 W	0257 2.5	0836 9.3	1508 2.8	2046 9.6		26 TH	0352 1.7	0928 10.0	1610 2.4	2146 10.0
12 TH	0331 2.8	0910 9.1	1542 3.2	2121 9.3		27 F	0435 2.2	1013 9.4	1652 3.0	2233 9.5
13 F	0404 3.1	0948 8.8	1617 3.5	2202 9.0		28 SA	0518 2.8	1059 8.9	1736 3.5	2322 9.0
14 SA	0442 3.4	1033 8.6	1659 3.7	2252 8.8		29 SU	0605 3.3	1151 8.5	1827 3.9	
15 SU	0527 3.6	1129 8.5	1754 3.9	2354 8.6		30 M	0018 8.6	0650 3.7	1252 8.2	1928 4.0

JULY

Day	Time/M	Time/M	Time/M	Time/M		Day	Time/M	Time/M	Time/M	Time/M
1 TU	0120 8.4	0801 3.8	1354 8.2	2030 3.9		16 W	0014 9.0	0653 3.4	1257 8.8	1933 3.5
2 W	0223 8.4	0900 3.7	1454 8.4	2128 3.7		17 TH	0128 8.9	0812 3.4	1415 8.9	2051 3.3
3 TH	0324 8.5	0953 3.5	1549 8.7	2220 3.3		18 F	0247 9.0	0929 3.1	1529 9.3	2204 2.9
4 F	0417 8.8	1042 3.2	1640 9.0	2311 3.0		19 SA	0402 9.4	1041 2.7	1637 9.8	2313 2.3
5 SA	0508 9.0	1130 3.0	1726 9.4	2358 2.7		20 SU	0509 9.9	1149 2.2	1737 10.4	
6 SU	0554 9.2	1218 2.8	1810 9.6			21 M	0019 1.8	0607 10.4	1249 1.8	1831 10.8 ○
7 M	0046 2.5	0638 9.4	1303 2.6	1849 9.8 ●		22 TU	0119 1.3	0700 10.7	1342 1.5	1920 11.0
8 TU	0131 2.3	0716 9.6	1344 2.4	1927 10.0		23 W	0209 1.0	0747 10.8	1430 1.4	2005 11.1
9 W	0211 2.2	0752 9.7	1423 2.4	2002 10.1		24 TH	0254 1.0	0829 10.7	1512 1.6	2046 10.9
10 TH	0247 2.2	0825 9.7	1458 2.4	2034 10.0		25 F	0334 1.3	0907 10.4	1549 2.0	2122 10.5
11 F	0321 2.3	0857 9.7	1531 2.6	2108 10.0		26 SA	0409 1.8	0943 9.9	1621 2.5	2159 10.0
12 SA	0352 2.5	0931 9.6	1603 2.8	2143 9.8		27 SU	0441 2.4	1017 9.4	1651 3.0	2235 9.4
13 SU	0423 2.7	1009 9.4	1638 3.0	2224 9.4		28 M	0512 3.0	1055 8.9	1725 3.5	2316 8.8
14 M	0459 3.0	1052 9.2	1720 3.3	2313 9.2		29 TU	0550 3.5	1140 8.4	1810 3.9	
15 TU	0547 3.2	1139 8.9	1818 3.5			30 W	0008 8.3	0641 4.0	1239 8.0	1913 4.2
						31 TH	0119 7.9	0748 4.2	1357 7.9	2029 4.2

AUGUST

Day	Time/M	Time/M	Time/M	Time/M		Day	Time/M	Time/M	Time/M	Time/M
1 F	0240 7.9	0903 4.1	1514 8.1	2142 3.9		16 SA	0237 8.4	0921 3.7	1525 8.8	2202 3.3
2 SA	0355 8.1	1009 3.9	1617 8.6	2245 3.5		17 SU	0404 9.0	1042 3.1	1637 9.5	2315 2.5
3 SU	0452 8.6	1108 3.4	1709 9.1	2342 3.1		18 M	0511 9.7	1147 2.4	1733 10.3	
4 M	0540 9.0	1200 3.0	1754 9.5			19 TU	0017 1.8	0603 10.4	1242 1.8	1822 10.9 ○
5 TU	0031 2.6	0622 9.5	1246 2.5	1834 10.0 ●		20 W	0109 1.2	0649 10.8	1330 1.4	1906 11.2
6 W	0114 2.2	0659 9.9	1327 2.2	1909 10.3		21 TH	0154 0.8	0728 11.0	1412 1.2	1944 11.3
7 TH	0154 1.9	0731 10.2	1405 1.9	1942 10.6		22 F	0233 0.8	0805 11.0	1447 1.3	2019 11.1
8 F	0229 1.7	0804 10.4	1440 1.8	2013 10.7		23 SA	0307 1.1	0837 10.7	1518 1.6	2051 10.8
9 SA	0300 1.7	0834 10.5	1511 1.9	2046 10.7		24 SU	0335 1.6	0907 10.3	1543 2.1	2121 10.3
10 SU	0331 1.8	0905 10.4	1542 2.1	2119 10.5		25 M	0359 2.2	0935 9.8	1607 2.6	2152 9.7
11 M	0400 2.1	0941 10.2	1613 2.5	2156 10.1		26 TU	0423 2.8	1004 9.3	1633 3.1	2224 9.0
12 TU	0433 2.5	1019 9.7	1652 2.9	2238 9.6		27 W	0452 3.4	1040 8.7	1709 3.7	2304 8.3
13 W	0515 3.1	1108 9.2	1744 3.4	2334 9.0		28 TH	0533 4.0	1126 8.1	1805 4.3	
14 TH	0615 3.6	1215 8.6	1900 3.8			29 F	0007 7.6	0639 4.5	1250 7.6	1931 4.6
15 F	0055 8.4	0745 3.9	1351 8.4	2034 3.8		30 SA	0202 7.3	0815 4.7	1446 7.7	2110 4.4
						31 SU	0338 7.7	0943 4.3	1559 8.3	2228 3.9

ENGLAND, SOUTH COAST

No.	PLACE	Lat. N.	Long. W.	TIME DIFFERENCES High Water (Zone G.M.T.)		Low Water		HEIGHT DIFFERENCES (IN METRES) MHWS	MHWN	MLWN	MLWS	M.L. Z_0 m.	
89	**DOVER**	(see page 22)		0000 and 1200	0600 and 1800	0100 and 1300	0700 and 1900	6·7	5·3	2·0	0·8		
85	Hastings .	50 51	0 35	0000	−0010	−0030	−0030	+0·8	+0·5	+0·1	−0·1	3·85	
86	Rye (Approaches) .	50 55	0 47	+0005	−0010	⊙	⊙	+1·0	+0·7	⊙	⊙	⊙	
86a	Rye (Harbour) .	50 56	0 46	+0005	−0010	⊙	⊙	−1·4	−1·7	§	§	1·97	
87	Dungeness .	50 54	0 58	−0010	−0015	−0020	−0010	+1·0	+0·6	+0·4	+0·1	4·13	
88	Folkestone .	51 05	1 12	−0020	−0005	−0010	−0010	+0·4	+0·4	0·0	−0·1	3·74	
89	**DOVER** . . .	51 07	1 19	STANDARD PORT				See Table V				3·70	
98	Deal .	51 13	1 25	+0010	+0020	+0010	+0005	−0·6	−0·3	0·0	0·0	3·54	
99	Richborough .	51 18	1 21	+0015	+0015	+0030	+0030	−3·4	−2·6	−1·7	−0·7	1·42	c
102	Ramsgate .	51 20	1 25	+0020	+0020	−0007	−0007	−1·8	−1·5	−0·8	−0·4	2·56	

FRANCE, NORTH COAST; CHANNEL ISLANDS

No.	PLACE	Lat. N.	Long. W.	TIME DIFFERENCES High Water (Zone −0100)		Low Water		HEIGHT DIFFERENCES (IN METRES) MHWS	MHWN	MLWN	MLWS	M.L. Z_0 m.	
1605	**ST. HELIER** . . .	(see page 206)		0300 and 1500	0900 and 2100	0200 and 1400	0900 and 2100	11·1	8·1	4·1	1·3		
	Channel Islands			(Zone G.M.T.)									
	Alderney												
1603	Braye .	49 43	2 12	+0050	+0040	+0025	+0105	−4·8	−3·4	−1·5	−0·5	3·62	
	Sark												
1603a	Maseline Pier .	49 26	2 21	+0005	+0015	+0005	+0010	−2·1	−1·5	−0·6	−0·3	4·87	
	Guernsey												
1604	St. Peter Port .	49 27	2 31	0000	+0012	−0008	+0002	−2·1	−1·4	−0·6	−0·3	4·99	
	Jersey												
1605	**ST. HELIER** . .	49 11	2 07	STANDARD PORT				See Table V				6·06	
1606	St. Catherine Bay .	49 13	2 01	0000	+0010	+0010	+0010	0·0	−0·1	0·0	+0·1	6·0	x
1607	Les Ecrehou .	49 17	1 56	+0004	+0012	+0010	+0020	−0·2	+0·3	−0·3	0·0	6·15	
1608	Les Minquiers .	48 58	2 08	+0007	0000	−0008	+0013	+0·4	+0·9	−0·1	+0·1	6·47	
1605	**ST. HELIER** . . .	(see page 206)		0100 and 1300	0800 and 2000	0200 and 1400	0700 and 1900	11·1	8·1	4·1	1·3		
	France			(Zone −0100)									
1609	Iles Chausey .	48 52	1 49	+0044	+0048	+0104	+0058	+1·8	+1·7	+0·8	+0·6	7·50	
1610	Dielette .	49 33	1 52	+0116	+0119	+0115	+0120	−1·6	−0·9	−0·5	−0·1	5·51	
1611	Carteret .	49 22	1 47	+0100	+0110	+0120	+0115	−0·1	+0·3	0·0	+0·1	6·30	
1612	Granville .	48 50	1 36	+0040	+0049	+0115	+0053	+1·7	+1·5	+0·5	+0·1	7·21	
1613	Cancale .	48 40	1 51	+0035	+0050	+0115	+0100	+2·2	+2·0	+1·0	+0·7	7·76	
1614	St. Malo .	F 48 38	2 02	+0034	+0044	+0105	+0050	+1·0	+1·0	+0·3	+0·1	6·85	
1615	Erquy .	48 38	2 28	+0030	+0040	+0035	+0032	+0·1	+0·4	0·0	0·0	6·40	
1616	Dahouet .	48 35	2 34	+0031	+0038	+0027	+0036	+0·1	+0·4	−0·2	−0·1	⊙	
1617	Le Légué .	48 32	2 44	+0030	+0045	+0035	+0031	+0·1	+0·4	0·0	0·0	5·6	x
1618	Binic .	48 36	2 49	+0030	+0045	+0035	+0031	+0·1	+0·4	0·0	0·0	5·6	x
1619	Portrieux .	48 38	2 49	+0030	+0045	+0030	+0030	+0·1	+0·4	0·0	0·0	6·38	
1620	Paimpol .	48 47	3 02	+0025	+0038	+0025	+0021	−0·8	−0·3	−0·9	−0·8	5·52	
1621	Ile de Bréhat .	48 51	3 00	+0020	+0040	+0010	+0015	−0·7	−0·1	−0·5	−0·2	5·85	
1622	Les Heaux de Brehat .	F 48 55	3 05	+0031	+0030	−0011	+0042	−1·3	−0·6	−0·7	−0·3	5·51	
1623	Lezardrieux .	48 47	3 06	+0026	+0038	+0015	+0020	−1·1	−0·6	−0·7	−0·4	5·57	
1624	Plougrescant .	48 51	3 13	−0005	+0004	−0017	−0013	−1·5	−0·7	−0·6	0·0	5·55	
1625	Tréguier .	48 47	3 13	−0004	+0012	−0018	−0007	−1·4	−0·7	−0·8	−0·4	5·46	
1626	Ploumanac'h .	48 50	3 29	0000	+0005	−0025	−0015	−2·2	−1·1	−0·7	−0·4	5·15	

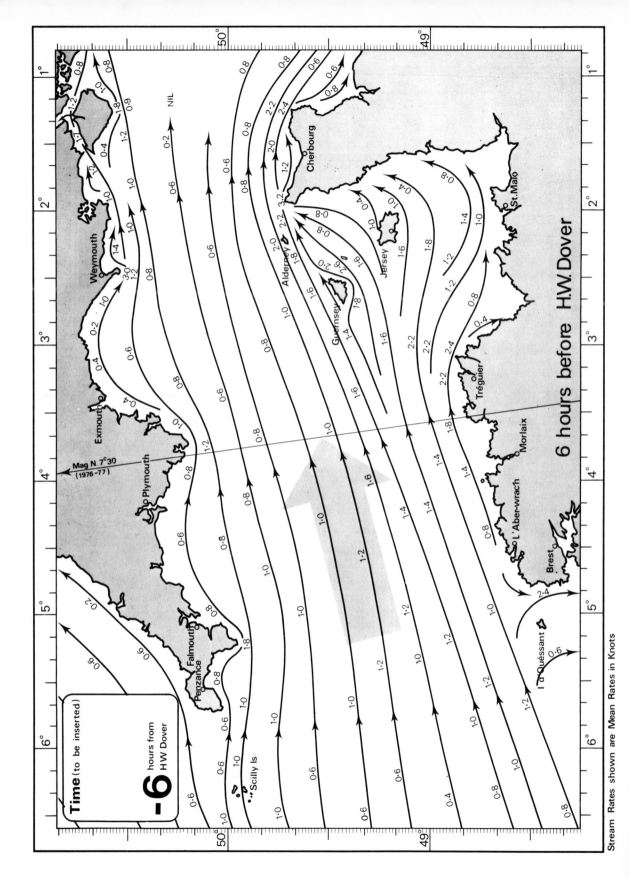

Time (to be inserted)

−6 hours from HW Dover

6 hours before H.W.Dover

Stream Rates shown are Mean Rates in Knots

112

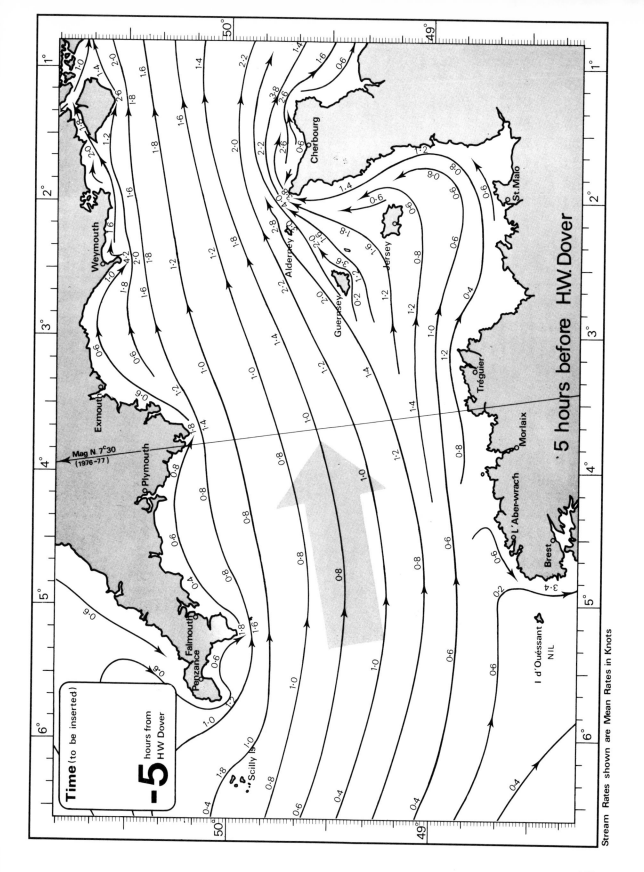

Stream Rates shown are Mean Rates in Knots

113

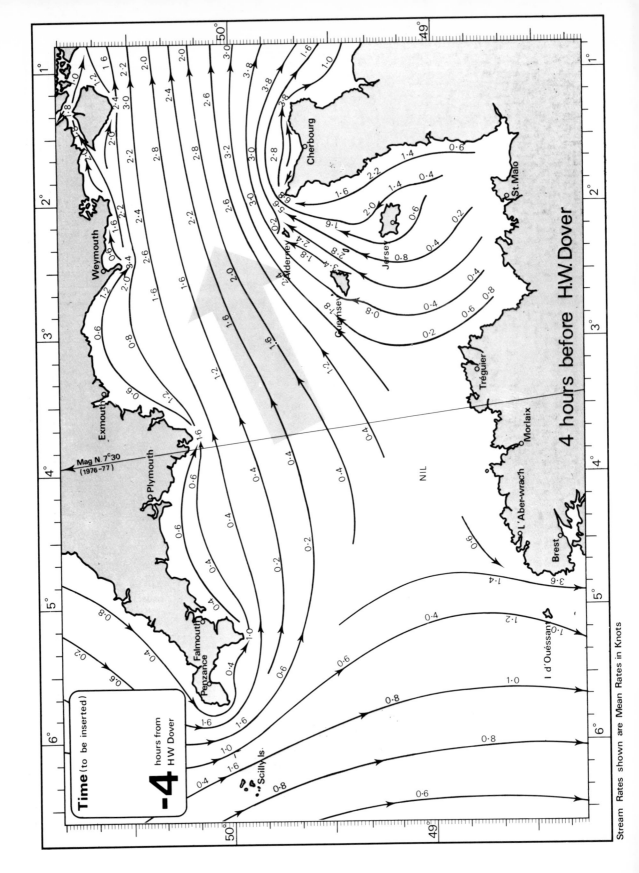

Stream Rates shown are Mean Rates in Knots

114

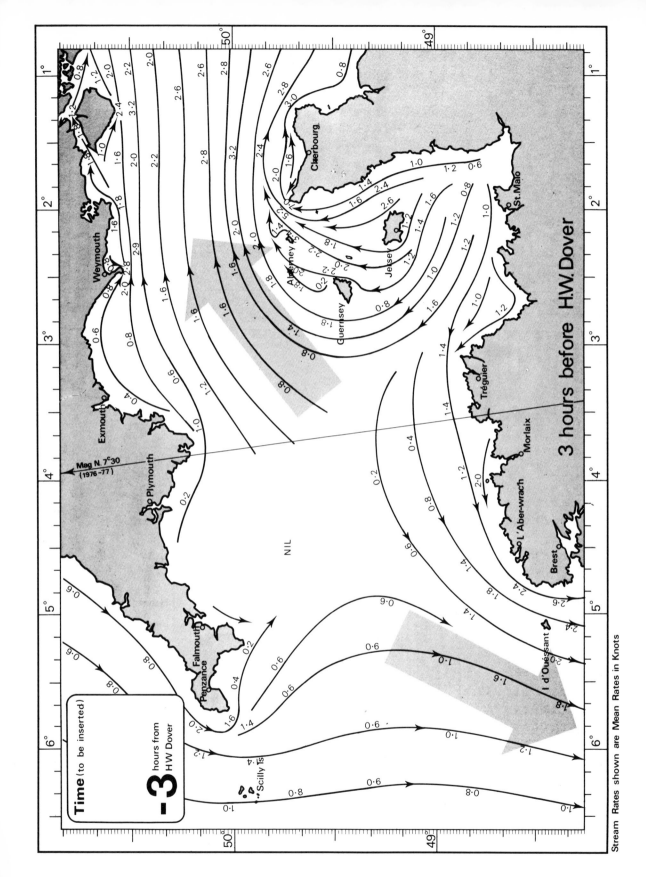

Time (to be inserted)

−3 hours from HW Dover

3 hours before H.W. Dover

Stream Rates shown are Mean Rates in Knots

115

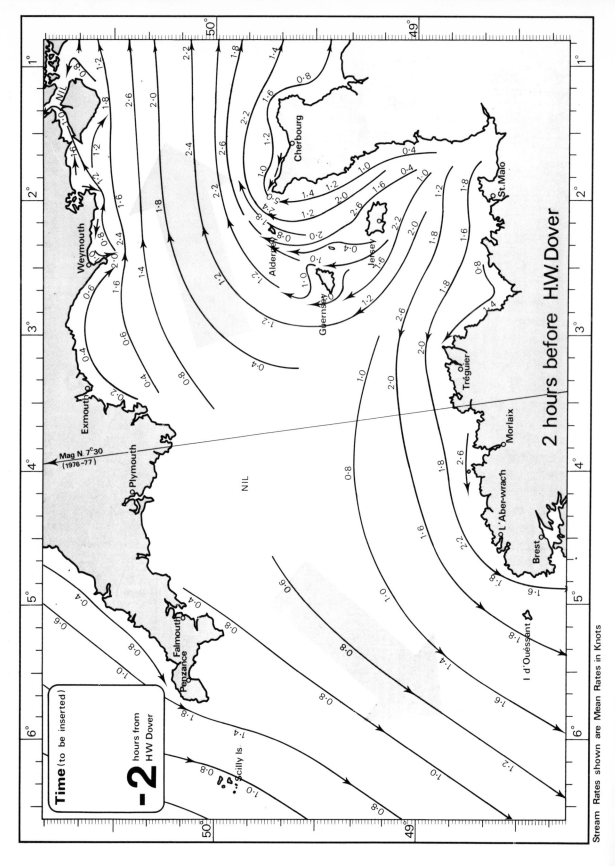

Time (to be inserted)

−2 hours from HW Dover

2 hours before H.W.Dover

Stream Rates shown are Mean Rates in Knots

116

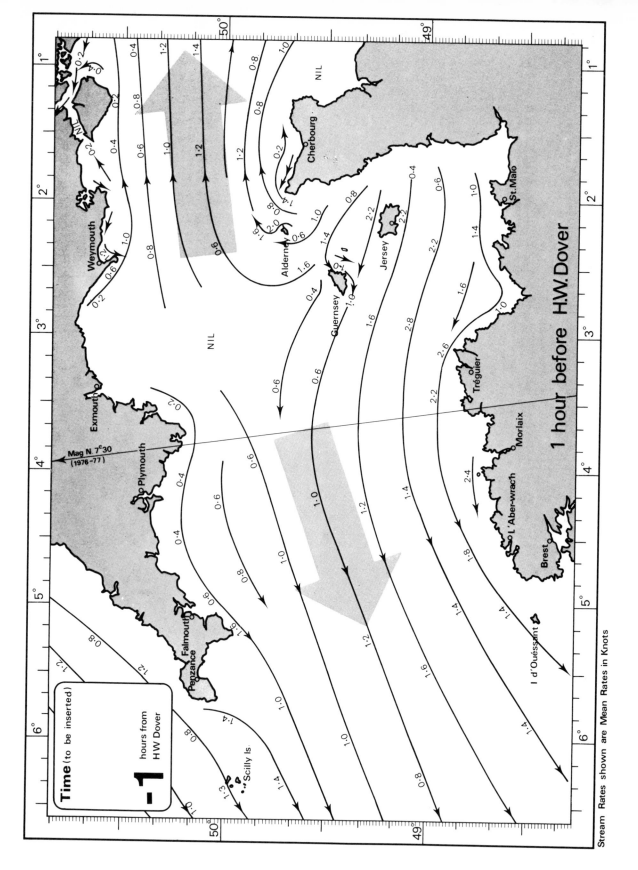

1 hour before H.W. Dover

Time (to be inserted)

-1 hours from HW Dover

Stream Rates shown are Mean Rates in Knots

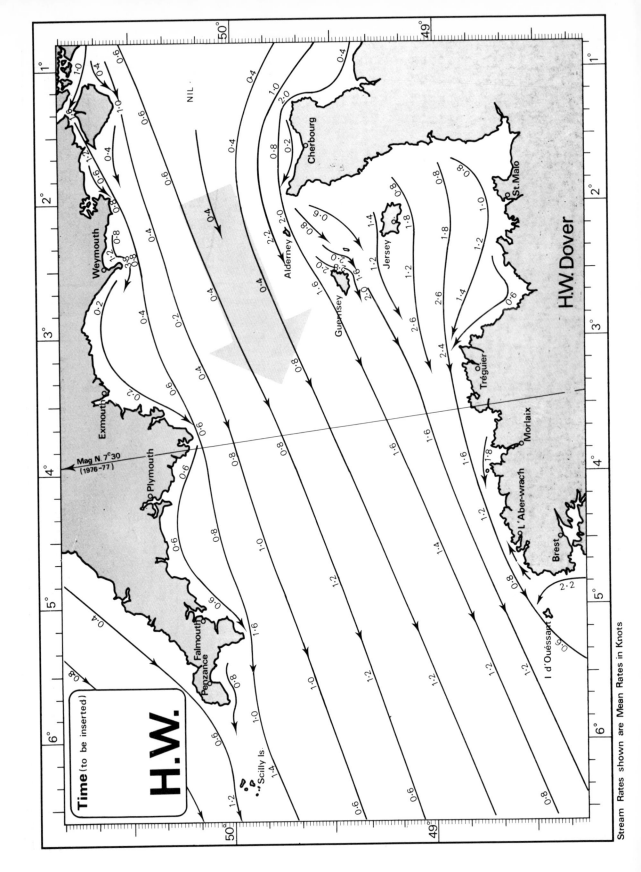

Stream Rates shown are Mean Rates in Knots

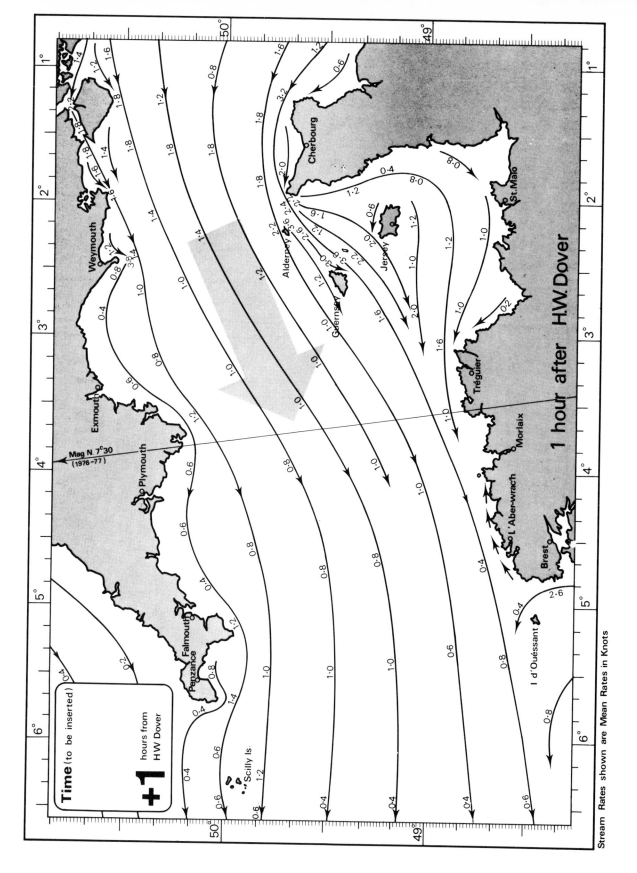

1 hour after H.W. Dover

Stream Rates shown are Mean Rates in Knots

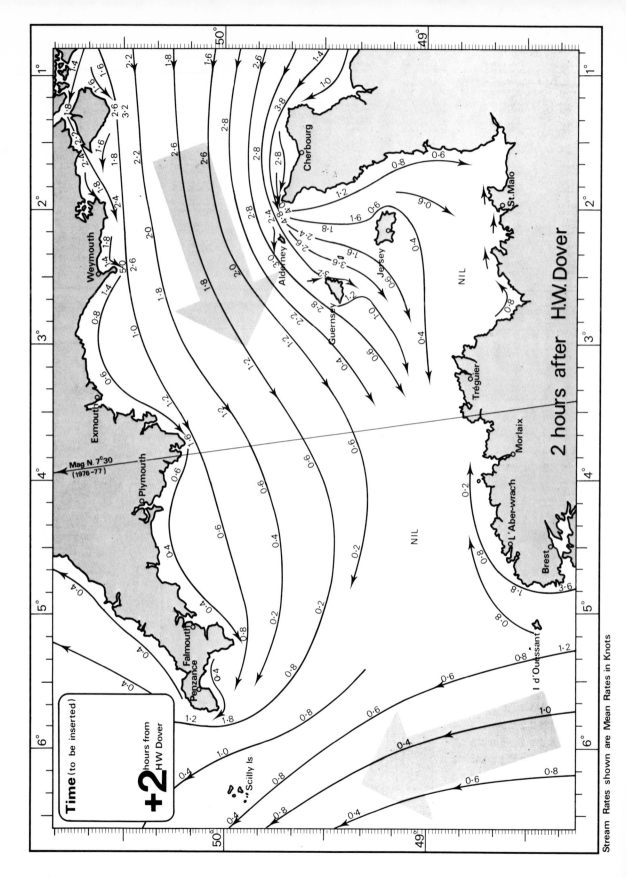

2 hours after H.W. Dover

Stream Rates shown are Mean Rates in Knots

Time (to be inserted)

+2 hours from HW Dover

Mag N. 7°30 (1976-77)

NIL

120

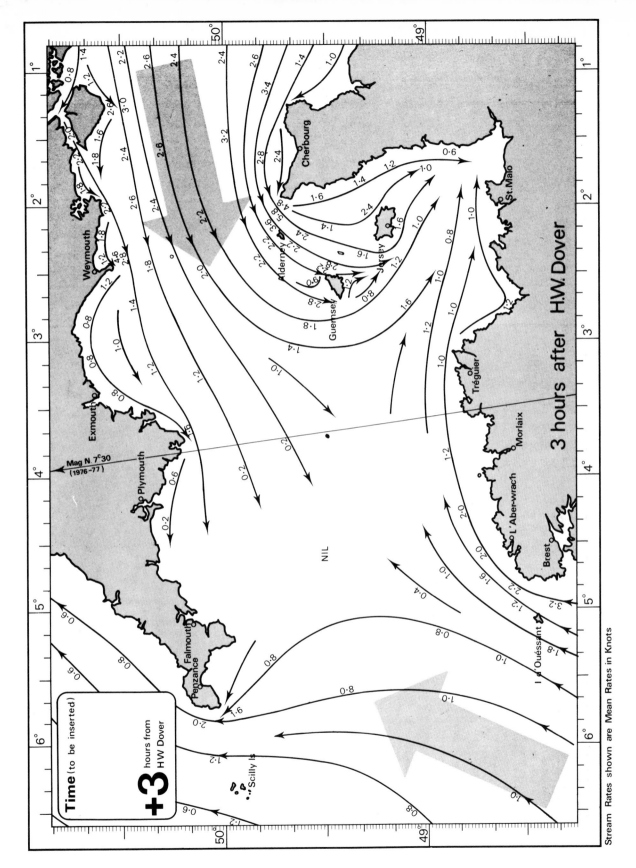

3 hours after H.W. Dover

Time (to be inserted)

+3 hours from HW Dover

Stream Rates shown are Mean Rates in Knots

121

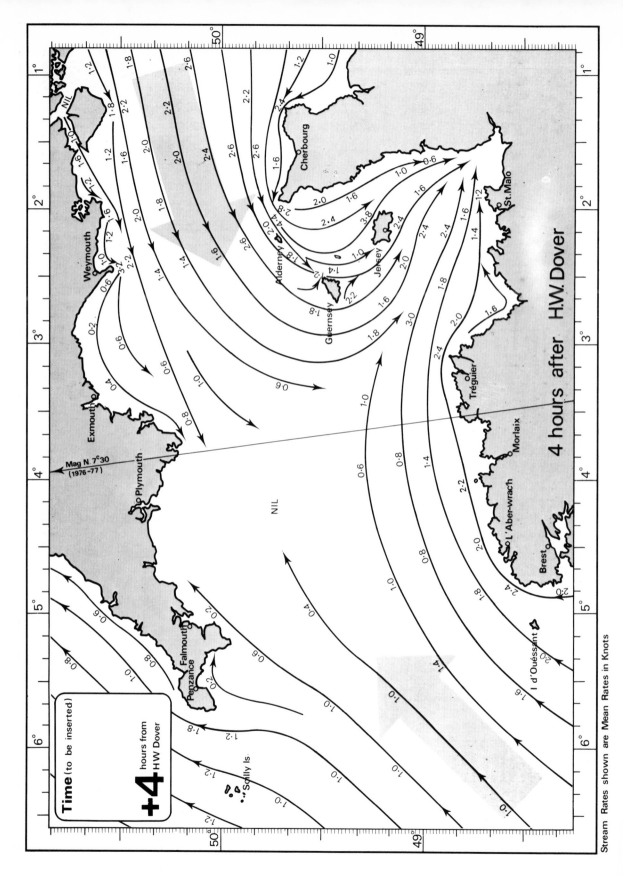

4 hours after H.W. Dover

Time (to be inserted)

+4 hours from HW Dover

Stream Rates shown are Mean Rates in Knots

122

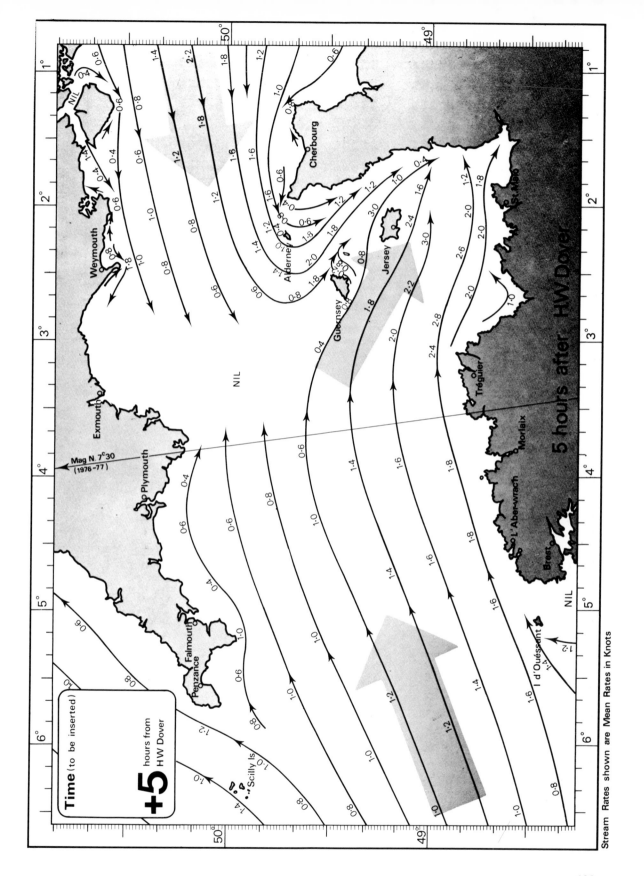

Stream Rates shown are Mean Rates in Knots

123

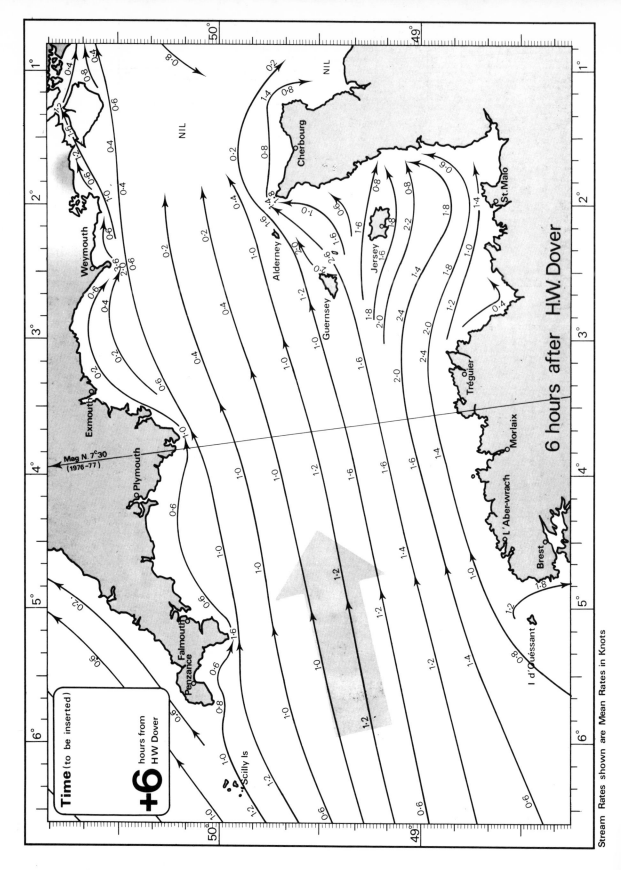

6 hours after H.W. Dover

Time (to be inserted)

+6 hours from H.W. Dover

Stream Rates shown are Mean Rates in Knots

124

KNOTS — Conversion from Mean Rate of Stream

Pencil-mark Height of H.W. Dover, read converted rate of stream from Column below mark

Mean Rate of Stream as Shown on Chart	5.0	5.2	5.4	5.6	5.8	6.0	6.2	6.4	6.6	6.8	7.0	7.2	7.4
0.2	0.1	0.1	0.2	0.2	0.2	0.2	0.2	0.2	0.3	0.3	0.3	0.3	
0.4	0.3	0.3	0.3	0.4	0.4	0.4	0.4	0.5	0.5	0.5	0.6	0.6	
0.6	0.4	0.4	0.5	0.5	0.6	0.6	0.7	0.7	0.8	0.8	0.9	0.9	
0.8	0.5	0.6	0.6	0.7	0.8	0.8	0.9	1.0	1.0	1.1	1.2	1.2	
1.0	0.6	0.7	0.8	0.9	1.0	1.0	1.1	1.2	1.3	1.4	1.4	1.5	
1.2	0.8	0.9	1.0	1.1	1.2	1.2	1.3	1.4	1.5	1.6	1.7	1.8	
1.4	0.9	1.0	1.1	1.2	1.3	1.5	1.6	1.7	1.8	1.9	2.0	2.1	
1.6	1.0	1.1	1.3	1.4	1.5	1.7	1.8	1.9	2.1	2.2	2.3	2.4	
1.8	1.1	1.3	1.4	1.6	1.7	1.9	2.0	2.2	2.3	2.5	2.6	2.7	
2.0	1.3	1.4	1.6	1.8	1.9	2.1	2.2	2.4	2.6	2.7	2.9	3.1	
2.2	1.4	1.6	1.8	1.9	2.1	2.3	2.5	2.6	2.8	3.0	3.2	3.4	
2.4	1.5	1.7	1.9	2.1	2.3	2.5	2.7	2.9	3.1	3.3	3.5	3.7	
2.6	1.7	1.9	2.1	2.3	2.5	2.7	2.9	3.1	3.3	3.5	3.8	4.0	
2.8	1.8	2.0	2.2	2.5	2.7	2.9	3.1	3.4	3.6	3.8	4.0	4.3	
3.0	1.9	2.2	2.4	2.6	2.9	3.1	3.4	3.6	3.8	4.1	4.3	4.6	
3.2	2.0	2.3	2.6	2.8	3.1	3.3	3.6	3.8	4.1	4.4	4.6	4.9	
3.4	2.2	2.4	2.7	3.0	3.3	3.5	3.8	4.1	4.4	4.6	4.9	5.2	
3.6	2.3	2.6	2.9	3.2	3.5	3.7	4.0	4.3	4.6	4.9	5.2	5.5	
3.8	2.4	2.7	3.0	3.3	3.6	4.0	4.3	4.6	4.9	5.2	5.5	5.8	
4.0	2.5	2.9	3.2	3.5	3.8	4.2	4.5	4.8	5.1	5.5	5.8	6.1	
4.2	2.7	3.0	3.4	3.7	4.0	4.4	4.7	5.0	5.4	5.7	6.1	6.4	
4.4	2.8	3.2	3.5	3.9	4.2	4.6	4.9	5.3	5.6	6.0	6.4	6.7	
4.6	2.9	3.3	3.7	4.0	4.4	4.8	5.2	5.5	5.9	6.3	6.6	7.0	
4.8	3.1	3.4	3.8	4.2	4.6	5.0	5.4	5.8	6.2	6.5	6.9	7.3	
5.0	3.2	3.6	4.0	4.4	4.8	5.2	5.6	6.0	6.4	6.8	7.2	7.6	
5.2	3.3	3.7	4.1	4.6	5.0	5.4	5.8	6.3	6.7	7.1	7.5	7.9	
5.4	3.4	3.9	4.3	4.7	5.2	5.6	6.1	6.5	6.9	7.4	7.8	8.2	
5.6	3.6	4.0	4.5	4.9	5.4	5.8	6.3	6.7	7.2	7.6	8.1	8.5	
5.8	3.7	4.2	4.6	5.1	5.6	6.0	6.5	7.0	7.4	7.9	8.4	8.8	
6.0	3.8	4.3	4.8	5.3	5.8	6.2	6.7	7.2	7.7	8.2	8.7	9.2	
6.2	3.9	4.4	4.9	5.4	5.9	6.5	7.0	7.5	8.0	8.5	9.0	9.5	
6.4	4.1	4.6	5.1	5.6	6.1	6.7	7.2	7.7	8.2	8.7	9.2	9.8	
6.6	4.2	4.7	5.3	5.8	6.3	6.9	7.4	7.9	8.5	9.0	9.5	10.1	
6.8	4.3	4.9	5.4	6.0	6.5	7.1	7.6	8.2	8.7	9.3	9.8	10.4	
7.0	4.5	5.0	5.6	6.2	6.7	7.3	7.8	8.4	9.0	9.5	10.1	10.7	

DECK LOG

Yacht **Zephyr**

Date 27.10.86

Passage **Needles to Aldemey**

Time	Log reading	Course ordered	Course steered	Wind Dir	Wind Force	Bar mB	Remarks
0300	39.6	210	212	W	6	987	0245 Reef in main Pumped bilges
0400	45.4	210	210	W	6/7	985	0330 Set No2 Jib
0415	46.9	210	212	W	7	985	0415 Aldemey lt 207° a/c 215° Cde la Hague 195°
0500	51.5	215	216	WNW	7	984	Rain squall – poor vis. 0500 Casquets RC 230°

Acknowledgments

Crown copyright. Reproduced from *Admiralty Tide Tables Vol 1, 1986* with the permission of the Controller of Her Majesty's Stationery Office.

Tidal predictions for Dover and St Helier have been computed by the Institute of Oceanographic Sciences: copyright reserved.